RENEWABLE ENERGY SOURCES IN REDUCING CARBON FOOTPRINT

RENEWABLE ENERGY SOURCES IN REDUCING CARBON FOOTPRINT

Edited by

Dr. Sujata Mustapure

Delve
PUBLISHING

www.delvepublishing.com

Renewable Energy Sources in Reducing Carbon Footprint

Dr. Sujata Mustapure

Delve Publishing
4164 Lakeshore Road
Burlington, ON L7L 1A4
Canada
www.delvepublishing.com

Email: orders@arclereducation.com

ISBN: 978-1-77956-422-1 (Hardcover)

ABOUT THE EDITOR

Dr. Sujata Nagnath Mustapure (1988) is presently serving as teaching associate, Department of Electrical & Other Energy sources, College of agricultural engineering & Technology, Vasantrao Naik Marathwada Krishi Vidhyapeet, Parbhani, Maharashtra. She obtained her B.Tech (agril. Engg.) in 2010 from Vasantrao Naik Marathwada Krishi Vidhyapeet, Parbhani. Gold medal in M.Tech (Agril. Engg.), 2014 from Dr. Panjabrao Deshmukh Krishi Vidyapeeth, Akola and Ph.D in Renewable energy engineering in 2019 from Maharana Pratap University of Agriculture and Technology, Udaipur. She has also worked as assistant professor in College of Agricultural, Naigoan, Vasantrao Naik Marathwada Krishi Vidhyapeet, Parbhani.

TABLE OF CONTENTS

LIST OF FIGURES

LIST OF TABLE

PREFACE

Renewable energy sources, such as electricity generated from wind, solar, and geothermal systems, as well as biomass energy conversion and modernized hydroelectric systems, are broadly classified as renewable. Biomass energy systems are diverse and include resources such as dedicated energy crops, wood waste, landfill gas, digester gas, animal waste, and municipal solid waste. However, definitions of renewable energy can differ between private entities and governmental organizations.

Since the Industrial Revolution, humanity has heavily exploited natural resources to fuel economic growth and societal development. This reliance has resulted in significant environmental impacts, including the widespread use of fossil fuels, deforestation, wetland drainage, coastal marine ecosystem degradation, and unsustainable land-use practices. These activities have led to increased anthropogenic emissions of climate-active gases, raising their concentrations in the atmosphere. Over the past 150 years, these processes are believed to have contributed to a rise of approximately 1°C in the global average near-surface temperature. One of the key challenges today is to mitigate these impacts by reducing anthropogenic carbon dioxide (CO_2) emissions and limiting the rise in global temperatures. Additionally, global production and trade dynamics play a crucial role in carbon emissions. Developed countries, which tend to import more carbon-intensive products while exporting fewer, effectively transfer carbon emissions to developing countries. This results in a net increase in carbon transfers from developing to developed nations through trade.

The textbook *Renewable Energy Sources in Reducing Carbon Footprint* offers a comprehensive exploration of various renewable energy sources and their impact on mitigating carbon emissions. Structured into nine detailed chapters, this book serves as an essential resource for understanding the role of renewable energy in combating climate change and reducing our carbon footprint.

The first chapter provides a foundational overview of renewable energy and the concept of carbon footprint. It introduces key principles and the significance of transitioning to renewable energy sources to address global environmental challenges. This chapter sets the stage for a deeper exploration of specific energy sources.

In the second chapter, the focus shifts to solar energy. It covers the fundamentals of solar power, including photovoltaic and solar thermal technologies, and examines their efficiency, applications, and potential for reducing carbon emissions.

The third chapter examines wind energy, discussing the mechanics of wind turbines, site selection, and the role of wind power in sustainable energy strategies. It highlights advancements in wind technology and their contributions to decreasing the carbon footprint.

Hydropower is explored in the fourth chapter, which examines various hydropower systems, including conventional dams and run-of-river systems. This chapter evaluates the environmental impacts, benefits, and challenges associated with hydropower as a renewable energy source.

The fifth chapter addresses biomass and bioenergy, detailing the different types of biomass resources, conversion technologies, and their role in reducing carbon emissions. It covers the use of dedicated energy crops, wood waste, and organic waste, emphasizing their potential and limitations.

Geothermal energy is the subject of the sixth chapter. It explores geothermal systems, including geothermal power plants and ground-source heat pumps, and assesses their efficiency and environmental benefits in reducing carbon emissions.

The seventh chapter covers ocean energy, including tidal, wave, and ocean thermal energy. It examines the technological developments and potential of ocean energy to contribute to global renewable energy goals.

Integration and storage of renewable energy are the focus of the eighth chapter. It addresses the challenges of integrating various renewable sources into the energy grid and the role of energy storage solutions in ensuring a stable and reliable energy supply.

The final chapter discusses the safety of energy sources and future trends and innovations. It reviews safety considerations for renewable energy technologies and explores emerging trends and innovations that could shape the future of renewable energy and further reduce carbon footprints.

This textbook is written for students, researchers, and professionals in the fields of environmental science, energy engineering, and sustainability. It is also a valuable resource for policymakers and industry stakeholders interested in understanding and implementing renewable energy solutions to combat climate change.

Chapter 1

Introduction to Renewable Energy and Carbon Footprint

Contents

INTRODUCTION

Renewable energy is essential for combating climate change and advancing environmental sustainability because it comes from natural processes that replenish themselves over time. Energy sources like solar, wind, hydro, geothermal, and biomass offer a clean substitute for fossil fuels, which are highly responsible for the emission of greenhouse gases. When burning, conventional energy sources release large amounts of carbon dioxide and other pollutants; in contrast, renewable energy technologies produce power with little to no emissions while in use. For example, wind turbines convert wind energy into electrical power and solar panels use sunlight to generate electricity; both have very low direct emissions.

The importance of renewable energy goes beyond its ability to lower emissions. Renewably sourced energy helps lessen dependency on carbon-intensive energy sources, which lowers total greenhouse gas emissions, by dispensing with the need for fossil fuel-based power generation. This change is essential to reducing the negative effects of climate change, such as extreme weather, rising sea levels, and ecosystem disruption, as well as to mitigate global warming. Furthermore, by diversifying energy sources and lowering reliance on imported fuels, the advancement and application of renewable energy technologies promote energy security. This improves resistance to changes in the market and interruptions in the supply chain.

Because it generates jobs in the manufacturing, installation, and maintenance sectors, renewable energy also promotes economic growth. Renewable energy sources become more widely available and affordable as costs come down, which encourages widespread adoption and further lowers carbon footprints. In general, renewable energy signifies a significant change toward a more resilient and sustainable energy system, which is necessary to meet long-term climate targets and protect the environment for coming generations.

1.1. OVERVIEW OF RENEWABLE ENERGY

Energy derived from naturally replenishing renewable resources within a human timescale is referred to as renewable energy. Sunlight, wind, rain, tides, waves, and geothermal heat are some of its sources. Certain renewable energy sources are not sustainable, despite the majority being. Certain biomass sources, for instance, are deemed unsustainable at the current rates of exploitation. Energy from renewable sources is frequently used for stand-alone power systems, air, and water heating and cooling, and electricity generation for the grid. Renewable energy sources account for about 20% of the total energy used by humans worldwide, including nearly 30% of electricity. Traditional biomass accounts for about 8% of energy consumption, although this is decreasing. Heat energy from contemporary renewable sources, like solar water heating, makes up more than 4% of energy consumption, while electricity makes up more than 6% (Figure 1.1).

Figure 1.1. *Renewable energy sources.*

Source: By Melanie Maecker-Tursun – Melanie Maecker-Tursun, www.ponymithorn.com, hello@ponymithorn.com, CC BY-SA 4.0, https://commons.wikimedia.org/w/index.php?curid=88679630.

The renewable energy sector employs over 10 million people worldwide, with solar photovoltaics being the industry leader in employment. Most newly installed electricity capacity worldwide is made up of renewable energy sources, which are also becoming more and more affordable. As a percentage of total energy consumption, renewable energy systems are contributing to worldwide energy consumption growth. The cheapest new-build electricity comes from onshore wind or photovoltaic solar in the majority of countries.

More than 20% of the energy supply in many countries around the world currently comes from renewable sources, with some producing more than half of their electricity this way. It is anticipated that national markets for renewable energy will keep expanding rapidly into the 2020s and beyond. A few nations use renewable energy sources to produce all of their electricity. Renewable energy resources are distributed across large geographic regions, while fossil fuels are concentrated in a small number of nations. Energy security, economic benefits, and climate change mitigation are all being significantly enhanced by the deployment of renewable energy and energy efficiency technologies. However, subsidies for fossil fuels worth hundreds of billions of dollars are impeding the use of renewable energy. International public opinion polls consistently show a strong preference for renewable energy sources like wind and solar energy. However, the International Energy Agency stated in 2021 that to achieve net zero carbon emissions, more work must be done to increase renewable energy, with generation needs to rise by roughly 12% annually through 2030.

Large-scale renewable energy technology projects are typical, but they are also appropriate for developing nations and rural, isolated places where energy is frequently essential to human development. Since the majority of renewable energy technologies generate electricity, further electrification is frequently implemented in conjunction with renewable energy sources. This approach has multiple advantages, including the ability to convert electricity into heat, mechanical energy with high efficiency, and clean energy at the point of consumption. Furthermore, electrification powered by renewable energy is more efficient, which results in a major decrease in the amount of primary energy needed. Nearly half of the increase in renewable electricity in 2021 came from China. Norway, which is renowned for producing hydroelectricity, used hydro energy for 45% of its total energy supply in 2021 (Figure 1.2).

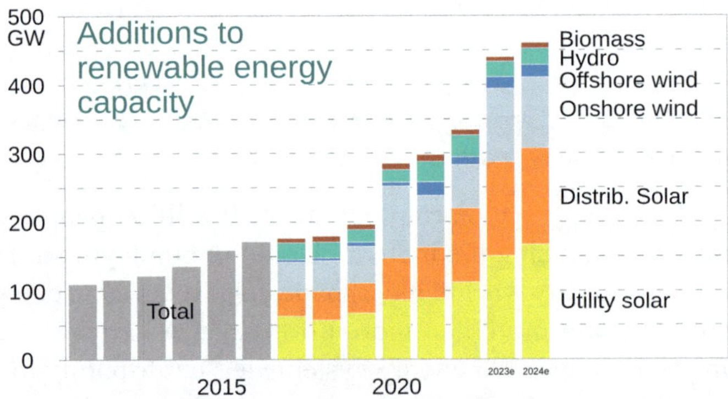

Figure 1.2. *Renewable energy capacity additions in 2020 expanded by more than 45% from 2019, including 90% more new wind power and a 23% expansion of new solar photovoltaic installations.*

Source: By RCraig09 – Own work, CC BY-SA 4.0, https://commons.wikimedia.org/w/index.php?curid=105192777.

Natural phenomena like sunlight, wind, tides, plant growth, and geothermal heat are all part of the flow of renewable energy. The International Energy

Agency states that renewable energy is produced by natural processes that are continuously replenished. It comes from the sun directly or from heat produced deep within the earth in all of its forms. Electricity and heat produced by solar, wind, ocean, hydropower, biomass, geothermal resources, biofuels, and hydrogen produced from renewable resources are all included in the definition (Figure 1.3).

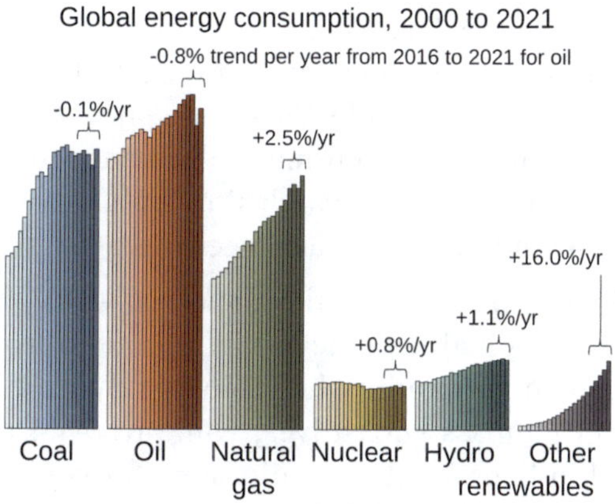

Figure 1.3. *Coal, oil, and natural gas remain the primary global energy sources even as renewables have begun rapidly increasing.*

Source: https://upload.wikimedia.org/wikipedia/commons/thumb/4/49/Global_Energy_Consumption.svg/1092px-Global_Energy_Consumption.svg.png.

In contrast to fossil fuels, which are depleted far faster than they are replenished, renewable energy sources. Unlike other energy sources, which are concentrated in a small number of countries, renewable energy resources and significant opportunities for energy efficiency exist over wide geographical areas. There would be major economic and energy security advantages from the quick adoption of energy efficiency and renewable energy, as well as from the technological diversification of energy sources. The cost of the technologies underlying many of these renewable energy sources, like wind and solar power, has decreased economically. In certain circumstances, switching to these sources will be less expensive than sticking with the present, ineffective fossil fuels. In addition, it would improve public health, lower premature mortality rates from pollution, and save associated health costs that could reach trillions of dollars annually. Environmental pollution, such as air pollution brought on by the burning of fossil fuels, would also be reduced. Quantifiable health benefits can considerably outweigh the costs of putting these decarbonization strategies into practice, according to several analyses of these strategies. It is anticipated that humanity will be able to obtain energy from renewable energy sources, such as wind and hydropower, for nearly another billion years. After that, an increase in solar heat is predicted, which will eventually cause the Earth's surface to become too hot for liquid water to exist.

Concerns about climate change and global warming, along with the ongoing decline in the price of some renewable energy equipment, like solar panels and wind turbines, are encouraging a rise in the use of renewables. Compared to many other sectors, this one fared better through the global financial crisis thanks to new government spending, regulations, and policies. However, as of 2019, the International Renewable Energy Agency states that to keep the rise in average global temperatures during the present century well below 2°C (3°F) over pre-industrial levels, renewables' overall share of the energy mix (including power, heat, and transportation) needs to grow six times faster.

A few million homes were powered by small solar PV systems as of 2011, and many more were supplied with electricity by micro-hydro systems set up as mini-grids. More than 166 million households rely on a new generation of more energy-efficient biomass cookstoves, while over 44 million households use biogas produced in household-scale digesters for cooking and/or lighting. According to Ban Ki-moon, the eighth Secretary-General of the United Nations, renewable energy has the power to propel the world's poorest countries to unprecedented heights of prosperity. At the national level, renewable energy already accounts for more than 20% of the energy supply in at least 30 countries worldwide. Even though many nations have different policy targets for longer-term shares of renewable energy, these are typically limited to the power sector, such as the European Union's 2030 target of 40% of all electricity generated.

Renewable energy often displaces conventional fuels in four areas: electricity generation, hot water/space heating, transportation, and rural (off-grid) energy services:

1. **Power Generation:** More than a quarter of electricity is generated from renewables.
2. **Heating and Cooling:** Solar water heating makes an important contribution to renewable heat in many countries, most notably in China, which now has 70% of the global total (180 GW^{th}). Most of these systems are installed on multi-family apartment buildings and meet a portion of the hot water needs of an estimated 50–60 million households in China. Worldwide, total installed solar water heating systems meet a portion of the water heating needs of over 70 million households. The use of biomass for heating continues to grow as well. In Sweden, the national use of biomass energy has surpassed that of oil. Heat pumps provide both heating and cooling, and also flatten the electric demand curve and are thus an increasing priority

Renewable thermal energy is also growing rapidly. About 10% of heating and cooling energy is from renewables.

3. **Transportation:** One of the efforts to decarbonize transportation is the increased use of electric vehicles (EVs). Despite that and the use of biofuels, such as bio jets, less than 4% of transport energy is from renewables (Armstrong and Hamrin; TERI, 2000; Moni, 2011).

1.1.1. HISTORY OF RENEWABLE ENERGY

Almost all energy was renewable until the mid-19th century when coal was discovered. More than a million years ago, people used traditional biomass as fuel for fires, which is the earliest known application of renewable energy. It took many hundreds of thousands of years before biomass was widely used for fire. Using wind to propel ships over water is arguably the second-oldest application of renewable energy. About 7000 years ago, ships in the Persian Gulf and on the Nile began engaging in this practice. Geothermal energy from hot springs has been utilized for space heating since the ancient Roman era and for bathing since Paleolithic times. When recorded history began, human labor, animal power, water power, wind, grain-crushing windmills, and firewood—traditional biomass—were the main sources of renewable energy. Werner von Siemens wrote the following in 1885 in response to the solid-state photovoltaic effect's discovery: In conclusion, I would say that, regardless of the discovery's immense scientific significance, its practical value will become clear when we consider that solar energy is a limitless and costless resource that will continue to shower down upon us for eons after the earth's coal reserves have been depleted and forgotten.

In the last few paragraphs of his 1905 book DieprotestantischeEthik und der Geist des Kapitalismus (The Protestant Ethic and the Spirit of Capitalism), Max Weber alluded to the demise of fossil fuel. Solar engine development went on until the start of World War I. In a 1911 Scientific American article, the significance of solar energy was acknowledged: when natural fuels run out in the far future, solar power will be the only thing keeping humans alive.

In 1956, the peak oil theory was published. The first wind turbines that produced electricity first appeared in the 1970s as a result of environmentalists pushing for the development of renewable energy as a means of escaping dependence on oil and a potential replacement for it. Solar energy has long been used for heating and cooling, but until 1980, the cost of solar panels prevented the construction of solar farms. Many regions of the world have switched from using fossil fuels to renewable energy sources since the turn of the twenty-first century.

1.2. RENEWABLE ENERGY SOURCES

Large, continuous, natural energy sources are referred to as renewable energy. This comprises energy from the sun, wind, water, geothermal sources, and some types of biofuels. The drawback of renewable energy is that significant infrastructure is typically needed to extract meaningful power. For instance, costly solar panels are needed to collect solar energy. Wind turbines are necessary for wind energy harvesting. Dams are needed to harvest energy from flowing water, and so forth. Renewable energy is underutilized today: less than 10% of the electricity in most nations comes from renewable sources.

In addition to being more readily available, fossil fuels currently yield a higher yield for every dollar spent on their extraction. Some nations, like Iceland and Norway, rely entirely on renewable energy sources to power their economies; however, this is largely due to their advantageous geographic location near geothermal hotspots. It will cost a lot of money to start the transition to renewable energy in other nations (Figure 1.4).

Figure 1.4. *Renewable energy sources are replenished naturally and are virtually inexhaustible. There are several sources of renewable energy that are available in the United States today including sunlight, wind, geothermal heat, flowing water, and biofuels.*

Source: By Warren Rohner from Cape Town, South Africa – Darling Wind Farm, CC BY-SA 2.0, https://commons.wikimedia.org/w/index.php?curid=15581862.

In the end, solar energy will be the renewable energy source with the most power. Millions of government and venture capital funds are being invested in startups aiming to lower the cost and increase the efficiency of solar cells, and progress is being made in this direction. Even more effectively than standard solar panels, solar furnaces use mirrors to concentrate the sun's rays into a focus, which is then used to boil liquid and power turbines. Since most of the ocean's surface is unused and there is very little life there, these areas are perfect for the installation of floating solar panels that will power the world's cities. Eventually, we'll take advantage of the enormous amounts of available space to install solar panels and direct energy where it's needed.

1.2.1. Types of Renewable Energy Sources

Different types of energies are considered renewable energies namely solar energy, wind energy, tidal energy, hydroelectric energy, geothermal energy, biomass energy, so forth.

1.2.1.1. Solar Energy

One of the most widely used and rapidly expanding renewable energy sources is solar energy. Technology has developed a method for connecting the energy of the sun through solar panels as a free renewable energy source. There are two types of solar panels: solar thermal panels and solar PV cells. Solar photovoltaic cells harvest solar radiation and convert it into electrical energy for use in various applications such as electric vehicles, appliances, and heating. Solar thermal panels harness solar radiation for use in showers, heating systems, faucets, and other fixtures. In the burgeoning market for renewable energy, solar energy is the best choice (Figure 1.5).

Figure 1.5. *Solar panel array.*

Source: By kallerna – Own work, CC BY-SA 4.0, https://commons.wikimedia.org/w/index.php?curid=103494974.

1.2.1.2. Biomass Energy

The most popular renewable energy source is biomass energy. It transforms organic materials, such as plants and animals, into another useful form of energy. For example, when plants use photosynthesis to absorb solar energy, this energy is then transferred through the plant's internal structure to produce biomass energy. For the purpose of producing biomass energy, wood, compost, and crops are the most common types. Inadequate management of biomass technology may result in detrimental consequences for the environment.

1.2.1.3. Wind Energy

Wind energy has been used for several years to power windmills, push sails, and also for generating force for water pumps. When contrasted to other types of renewable energies, wind energy is considered steady as well as very reliable.

1.2.1.4. Hydroelectric Energy

The hydroelectric energy produces electricity by rotating turbines powered by the flow of water (Figure 1.6). 20% of the world's energy needs are met by this renewable energy, according to the US Geological Survey. There are certain problems with hydroelectric power. This energy can be produced by the dammed rivers; if not, it may significantly impact wildlife, the soil, and fish communities that must pass through the river dams.

Figure 1.6. *A hydroelectric dam slows the flow of a river in China to generate power.*

Source: By Source file: Le Grand PortageDerivative work: Rehman – File:Three_Gorges_Dam,_Yangtze_River,_China.jpg, CC BY 2.0, https://commons.wikimedia.org/w/index.php?curid=11425004.

1.2.1.5. Tidal Energy

Tidal energy is the same as wind energy but these are predictable as well as steady. This is the main reason that tidal energy sources are called potential sources. Tidal mills have been used since the ancient days to the middle ages similar to windmills.

Usually, tidal energy has faced relatively high costs as well as incomplete accessibility of sites through suitably high tidal ranges. However, several current technological developments both in technology and design point out that the entire tidal power availability may be superior to previous, and the environmental costs may be getting down to competitive stages.

1.2.1.6. Geothermal Energy

The word "geothermal" comes from the Greek word "geo," which means "earth." It absorbs heat from the planet and transforms it into energy. Examples of earth-generated energy that can be used to generate energy include steam energy and hot water. Because the Earth's water is regularly replenished by rainfall and the heat required is produced by the planet itself, it is referred to as a renewable energy source (Figure 1.7).

Figure 1.7. *A geothermal power station in the Philippines.*

Source: By Mike Gonzalez (TheCoffee) – Own work, CC BY-SA 3.0, https://commons.wikimedia.org/w/index.php?curid=1246507.

Ground-based heat pumps can be fixed to connect the normal heat from underground using fluid tubes covered outside the assets. The fluid in the tubes absorbs the heat from the ground so it can be used to heat your home and water. For assets that are located close to a river or lake, it is achievable to fix a heat pump for a water source. These pipes are flooded in the water as well as heat pump drives a heat that absorbs liquid during the arrangement of piping. This liquid removes normal heat from the nearby water to be utilized in the seating arrangement (Armstrong and Hamrin; TERI, 2000; Moni, 2011).

1.2.2. Importance of Renewable Resources of Energy

Renewable energy is abundant, and dependable, and could become extremely affordable as infrastructure and technology advance. Together with biofuels that are produced and harvested without the use of fossil fuels, they comprise solar, wind, geothermal, hydropower, and tidal energy (.). Non-renewable energy sources, like coal and petroleum, will only get more expensive as supplies run out and demand rises. These resources require expensive explorations as well as potentially hazardous mining and drilling. Because renewable energy emits very little carbon dioxide, it contributes to the fight against climate change brought on by the burning of fossil fuels.

Figure 1.8. *Imagine powering civilization entirely with energy from renewable sources: wind, sun, water (hydroelectricity), naturally occurring heat (geothermal), and plants.*

Source: By Thomas Lloyd Group – Own work, CC BY-SA 4.0, https://commons.wikimedia.org/w/index.php?curid=77438927.

1.2.2.1. Renewable How?

The reason renewable energy sources get their name is that, except for tidal and geothermal energy, sunlight is the primary source of replenishment. Wind is a result of uneven surface heating by the sun. The water cycle is also fueled by sunlight and is controlled by hydropower, which includes less intrusive methods like hydroelectric dams and systems that use streams or ocean currents. Light is used to grow biofuels. Since geothermal energy is produced by radioactive decay in the Earth's core, which isn't predicted to cool down anytime soon, it is regarded as renewable energy. The tides are brought on by the sun and moon's gravitational pull.

1.2.2.2. Accessibility

The reserves of oil, natural gas, and coal are limited and hidden. Each resource is submerged beneath the surface of the earth or in the ocean in an undetermined and finite amount. Finding new sources gets harder and more expensive as more are harvested, and using them gets trickier and occasionally riskier as well. Large volumes of natural gas must be burned to refine marginal reserves, like oil sands, into usable oil 1. Submarine drilling can result in disastrous mishaps, like the 2010 British Petroleum oil spill 2. In contrast, renewable energy can be found just as easily as sunlight or wind.

1.2.2.3. Reliability, Stability, and Safety

Numerous factors, including political stability in historically unstable regions, influence the daily price of oil. Conflicts over politics have contributed to energy crises, such as those that struck in 1973 and 1979. Since renewable energy can be produced locally, it is not susceptible to political upheavals in other places. With renewable energy, many of the safety issues associated with fossil fuels—such as explosions on oil platforms and collapsing coal mines—are eliminated.

1.2.2.4. Pollution

Fossil fuels are significantly dirtier than renewable energy. Mercury and other heavy metals are among the solid toxic wastes produced by coal mining, petroleum exploration, and refinement. In addition to consuming a lot of water, burning coal to generate electricity frequently releases lead and arsenic into surface waters and emits sulfur dioxide, carbon dioxide, nitrogen oxides, and mercury into the atmosphere. Petroleum products, including gasoline, also contribute to pollution. These pollutants deplete the ozone layer, produce acid rain that damages buildings and destroys delicate ecosystems, and cause respiratory illnesses and deaths in humans.

1.2.2.5. Climate Change

Strong consensus in the scientific community states that climate change and global warming are occurring and are caused by human production of carbon dioxide and other greenhouse gases. Climate change may also damage agriculture, cause widespread extinctions, imperil clean water supplies and aid the spread of tropical diseases (Armstrong and Hamrin; TERI, 2000; Moni, 2011).

1.2.3. Advantages of Renewable Energy

1.2.3.1. Renewable Energy Is Eco-Friendly

It emits little to no greenhouse gas and has low carbon emissions, making it a clean energy source. High emissions of greenhouse gases, primarily carbon dioxide, from fossil fuels are largely to blame for climate change, global warming, and deteriorating air quality (Figure 1.9). Acid rain is caused by sulfur emissions into the atmosphere, which is also a result of fossil fuels. Structures may sustain damage from acid rain. Because they release no harmful gases into the atmosphere, solar, and wind power are regarded as environmentally benign. Utilizing renewable energy sources significantly lowers reliance on fossil fuels as a source of energy, which lowers air pollution.

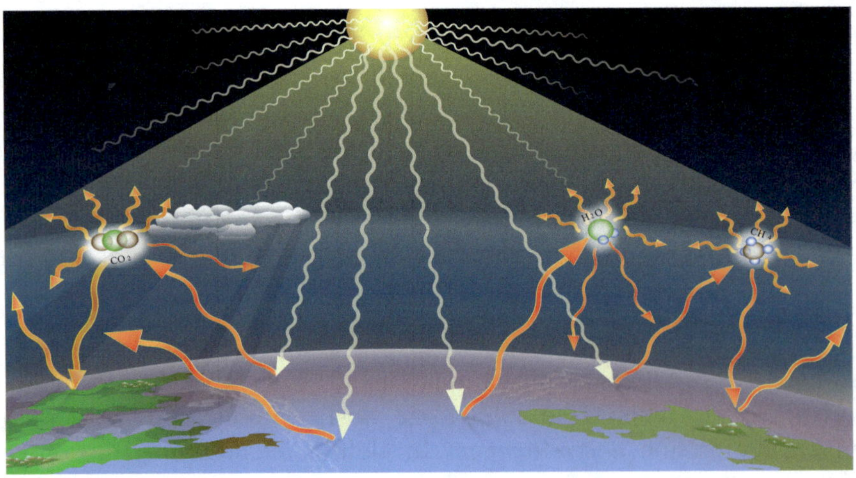

Figure 1.9. *The greenhouse effect of solar radiation on the Earth's surface caused by greenhouse gases.*

Source: https://upload.wikimedia.org/wikipedia/commons/0/04/Greenhouse-effect-t2.svg.

1.2.3.2. It's a Renewable Resource

This suggests that they are a sustainable source of energy and that there is no chance that they will run out over the course of a lifetime. Energy sources like fossil fuels, which include coal, gas, and oil, are regarded as finite resources with a high likelihood of depletion. Developing nations can benefit from less reliance on fossil fuels thanks to renewable energy. Strong winds, heat rising from the earth's surface, sunlight, and flowing water can all provide a country with a large and consistent supply of energy for a long time.

1.2.3.3. Renewable Energy Is a Reliable Source of Energy

The use of fossil fuels has skyrocketed over the last few decades. Because of our over-reliance on fossil fuels, our security is now in jeopardy. Fossil fuels are prone to energy price spikes, trade disputes, political unrest, and needless wars. These factors have a substantial impact on a country's economy in addition to a country's energy policies.

The majority of people contend that wind and solar energy are unstable, but a strong infrastructure disproves this claim. There can be very few interruptions to the production of electricity if solar and wind power plants are dispersed over a wide geographic area. This is because weather disruptions in one place cannot be replicated in another.

1.2.3.4. Leads to Job Creation

Since renewable energy is less expensive than the majority of conventional energy sources, it makes perfect financial sense. Most global economies have seen an increase in stable, new jobs since the advent of renewable energy. For example, the UK and Germany's persistent efforts to promote and develop the use of renewable energy sources have already resulted in the creation of numerous jobs in those countries.

1.2.3.5. Renewable Energy Has Stabilized Global Energy Prices

Switching to renewable energy sources means global energy prices will remain stable. This is because, unlike fossil fuels, which fluctuate in price based on the level of inflation at the time and the resource's availability, the cost of renewable energy is determined by the initial cost of installing renewable energy technologies. The initial expenses would be the only thing that the respective governments would have to cover (Figure 1.10).

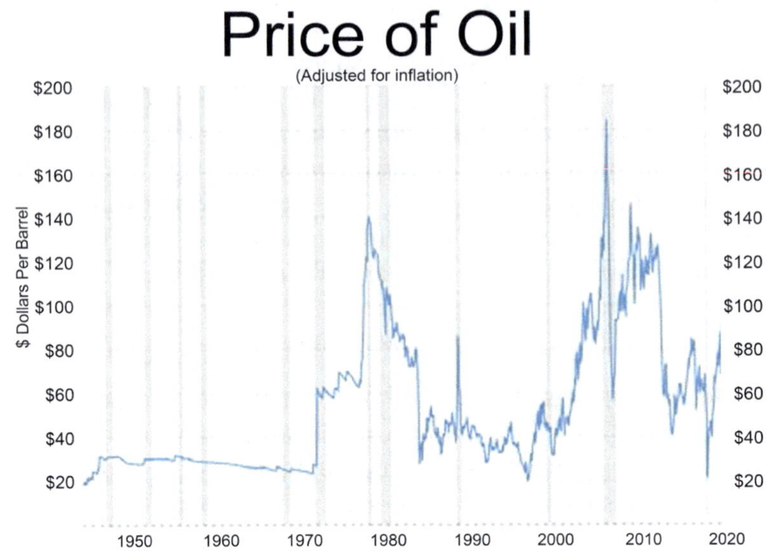

Figure 1.10. *Oil price shock – Big changes four years later.*

Source: By Macrotrends – https://www.macrotrends.net/1369/crude-oil-price-history-chart, Public Domain, https://commons.wikimedia.org/w/index.php?curid=117009809.

1.2.3.6. Less Maintenance of Facilities

Once the infrastructure for the harnessing of renewable resources is laid down, there is low to zero maintenance required. This means that the owners of the facilities will reap big profits while providing cheap electricity to the population.

1.2.3.7. Boosts Public Health

This is an obvious choice. Governments could improve the health of the populace by constructing more renewable energy facilities. According to a study conducted by the U.S. According to Environmental Protection Agency, Americans spend between $361 and $886 billion annually on population health. A significant portion of this budget is used to prevent and treat diseases like cancer, neurological disorders, and heart disease that are linked to the use of fossil fuels.

Greenhouse, carbon, and sulfur compounds emitted by fossil fuels are risky to our health if inhaled over time. This is reason enough to consider renewable energy moving forward.

1.2.3.8. Empowering People in the Countryside

The majority of renewable energy generation occurs in isolated areas. This implies that nearby towns would receive a fair portion of the electricity produced, which would eventually spark the social and economic revitalization of those impoverished areas. Untapped development opportunities will arise from the electrification of those areas, including the development of geothermal greenhouses and district heating of towns and communities using hot water produced by the energy extraction of forestry and agricultural biomass (Armstrong and Hamrin; TERI, 2000; Moni, 2011).

1.2.4. Disadvantages of Renewable Energy

1.2.4.1. The Electricity Generation Capacity is Still Not Large Enough

There are still challenges to the generation of large quantities of power in renewable energy technology compared to traditional forms of energy generation like fossil fuel. Fossil fuel still produces large quantities of electricity today, by far. This, essentially, means that it can't be solely relied upon to power the whole nation.

This means that either we need to set up more such facilities to match up with the growing demand or look for ways to reduce our energy consumption. This phenomenon indicates that a balance of different energy sources will still prevail for some years to come.

1.2.4.2. Renewable Energy Can Be Unreliable

Renewable energy technologies totally depend on the weather (e.g., sun, and wind) to be able to harness any energy. In case atmospheric conditions are not good enough, renewable energy technologies would lack the ability to generate electricity. This might instigate campaigns by the authorities to reduce energy usage to serve the population for a longer period.

1.2.4.3. Low-Efficiency Levels

Renewable energy technologies are still significantly new to the market, meaning, they still lack the much-needed efficiency. This poses forecast problems and investors may shy away from investing their money for fear of not getting returns pretty quickly.

1.2.4.4. Requires a Huge Upfront Capital Outlay

Setting up renewable energy generation facilities requires a huge financial outlay. Installation of wind turbines, solar panels, and hydroelectricity plants is relatively expensive. These plants require upfront investments to build, have high maintenance expenses and require careful planning and implementation. Also, the electricity generated needs to be delivered to towns and cities, which means the additional cost of installing power lines (Armstrong and Hamrin; TERI, 2000; Moni, 2011).

1.3. AVAILABILITY OF ALTERNATIVE RESOURCES

Although energy cannot be created, it can be transformed via technological, biological, and chemical processes. For example, biomass can be used to produce electricity, heat, and biofuels, and solar and wind energy can be converted into power and heat. The good news is that we already possess all the energy resources and conversion technologies required to guarantee a steady supply of clean, green energy.

Our planet has the opportunity to lower carbon emissions, purify the air, and establish a more sustainable foundation for human civilization through the use of renewable energy. Energy from renewable sources is crucial to a comprehensive plan for sustainable development. They contribute to lessening reliance on energy imports, protecting the environment and guaranteeing a sustainable supply. Furthermore, the long-term competitiveness of industries as well as regional development and employment can be positively impacted by renewable energy sources. A more varied, stable, and balanced pool of energy sources will be made available by renewable energy sources. Given the speed at which Brazil, China, and India are growing as well as the rest of the world, the question now is not if we will increase our use of renewable energy sources, but rather how quickly.

The technically exploitable amounts of energy in the form of electricity, heat, and chemical energy from renewable sources exceed the current world energy consumption by about sixfold.

The transition into distributed and decentralized renewable energy systems has to be associated with multiple measures:

1. Renewable energies should remain the priority (solar, wind, hydro, biomass, geothermal).
2. Improving energy efficiency.
3. Construction of smart grids.

4. Creating power storage facilities.
5. Future oriented and innovative policy in national, regional, and global contexts.
6. Creating a global climate framework.
7. Intensifying research and education activities.
8. Improving the cooperation between nations as well as between public and private sectors.

1.3.1. Power Plants

Power plants have typically been sizable, centralized structures (Figure 1.11). A new trend in energy generation is distributed generation, where larger units are being replaced by smaller ones and energy conversion units are located close to energy consumers. An eco-friendly, dependable, and efficient replacement for the conventional energy system is a distributed energy system.

Figure 1.11. *Nuclear power plant operators worldwide place their trust in KSB pumps.*

Source: https://en.wikipedia.org/wiki/File:2011-05-10_18-57-46_Switzerland_-_Wil_crop.jpg.

Distribution is the process of using low-voltage distribution lines to deliver electricity to a retail customer's residence or place of business. On-site generation, dispersed generation, embedded generation, decentralized generation, and on-site generation are other terms for distributed generation. Distributed energy, also known as decentralized energy, uses numerous tiny energy sources to produce fuels, heat, and electricity.

Because the electricity is generated close to its point of use, possibly even within the same building, less energy is lost during transmission. This lessens the size and quantity of power lines that need to be built as well. Reducing the amount of electricity used (through load management, energy conservation, etc.) and supply produced at or close to the location of power consumption. To meet local (substation level) peak loads and/or eliminate the need to construct new (or upgrade existing) local distribution lines, distributed generation systems place generation on a utility's distribution system.

1.3.2. Distributed Energy Generation

Distributed generation is also defined as the installation and operation of small modular power generating technologies that can be combined with energy management and storage systems. It is used to improve the operations of the electricity delivery systems at or near the end user. These systems may or may not be connected to the electric grid.

1.3.2.1. Energy Supply

In a Feed-in Tariff (FIT) program, distributed power sources typically have low maintenance costs, low emissions, and high efficiency. These characteristics used to be necessary for large, complex plants with specialized operating engineers to reduce pollution. Modern embedded systems, on the other hand, can offer these features in conjunction with renewable energy sources like geothermal, wind, and solar power. This lowers the maximum capacity of a power plant that can turn a profit (Figure 1.12).

There won't be any sizable fossil fuel-based power plants or sizable high-voltage transmission lines when designing a new power, heating, or transportation fuels system on paper. Every community within this new energy supply structure will have access to a range of locally available energy sources, including biomass, solar, wind, and other sources.

There will be two main sources of supply: wind and solar. For supplemental power and heat in situations where solar and wind energy are insufficient, biomass will be particularly significant. It is essential to balance the varying power generated by solar and wind; thermal and chemical storage solutions will be used. Future supply networks may come in a variety of sizes. The largest will apply to an area or city, and the smallest will be for a single-family home or settlement.

Figure 1.12. *Electricity station for energy supply. Free vector illustration for design.*

Source: By Dirk Jansen, CC BY 2.0, https://commons.wikimedia.org/w/index.php?curid=63864939.

As small scale technologies can be mass produced and thereby cheap, it may well prove preferable to and most economical to divide up cities in many independent and autonomous decentralized systems. In this section, we will call them off-grid as there is no advantage to have international interregional grid structures. So far off-grid supply has referred to unserved areas in developing countries without a national grid.

Distributed energy resource systems are small-scale power generation technologies (typically in the range of 3 kW to 10,000 kW) used to provide an alternative to or an enhancement of the traditional electric power system.

Due to the decentralized character of renewable energies, the future rational option will be to apply off-grid technologies to all types of communities, urban, and rural in fossil fuel served communities in industrialized countries and for unserved regions in developing countries. As the cost of fossil fuels increases and of renewable energy technologies declines and government support increases, investors, utilities, and governments are exploring ways to implement or invest in distributed renewable energy programs, projects, and companies.

1.3.2.2. Community Power

The word "community" can mean different things to different people. A community is any size social group whose members reside in a particular area. Accordingly, the phrase refers to "communities of locality," or places that are close together geographically, like a town, city, neighborhood, or district. The publication focuses on energy that is produced locally and dispersed to users within a specific area. A household's energy production can range from that of several buildings to that of a larger community through distributed energy generation. At least some of the total energy produced should ideally be distributed and used locally, though some energy may be fed back into the electrical grid (Figure 1.13).

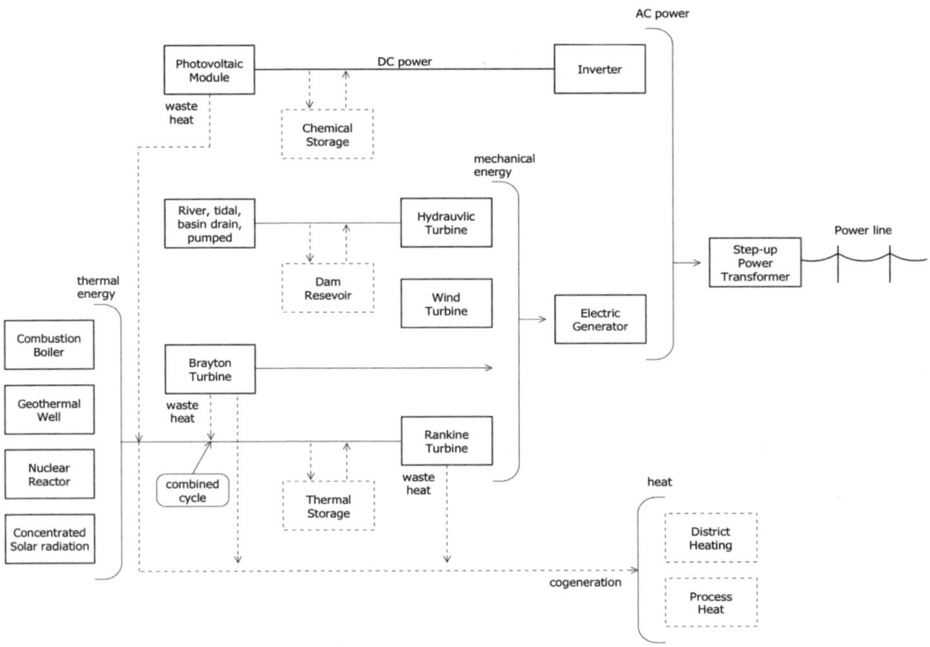

Figure 1.13. *Modular blocks.*

Source: By Robertiki – Own work, CC BY-SA 4.0, https://commons.wikimedia.org/w/index.php?curid=64195506.

1.3.2.3. Off-Grid Systems

An autonomously controlled power supply that is at least as dependable and high-quality as the public power grid is offered by off-grid systems. The phrase "off-grid" describes a situation where one is not connected to any kind of grid, primarily the national or main electrical transmission grid. Off-grid electrification is a method of obtaining electricity that is utilized in nations and regions where there is limited access to it because of dispersed or remote populations. It could be any form of energy production. With a generator and sufficient fuel supplies, electricity can be produced locally using renewable energy sources like solar, wind, or geothermal energy (Figure 1.14).

Figure 1.14. *To make the right choice it is important to comprehend the key differences between off-grid and on-grid solar systems.*

Source: By WiNG – Own work, CC BY-SA 3.0, https://commons.wikimedia.org/w/index.php?curid=7155522.

It can also be connected to local and national grids to substitute energy supply generated by nuclear or other non-renewable fuel sources, which is being called green electricity in some industrialized countries. The future of off-grid electricity supply since having global, interregional grid structures has no benefits. Up until now, off-grid supply has only been associated with underdeveloped nations without a national grid. Because renewable energy sources are decentralized, off-grid technologies will be applied to all kinds of communities—rural and urban—in areas that are served by fossil fuels in industrialized nations as well as to underserved areas in developing nations. The shift to decentralized energy sources may cause what was previously an exception to become the norm. Comparable deep shifts have frequently been observed in the long historical perspective. It's time to get ready for the end of the fossil fuel era because of the impending climate change and the immensity of the risks involved (Armstrong and Hamrin; TERI, 2000; Moni, 2011).

1.3.3. Concept of FAO, UN-Integrated Energy Communities (IEC)

Communities could benefit from renewable energy not just in the obvious sense but also in terms of improved development opportunities. There is a huge unmet need for affordable, off-grid, small-scale solutions to meet people's diverse energy needs. Individuals in developing nations are acutely aware of this. Offering them new choices that would genuinely satisfy their requirements and involve them in determining and organizing their own supply could make the success of offering renewable energy services a reality.

As long as energy is distributed fairly to those living in underserved areas, it is one of the key components of human empowerment. As a result, everyone should be entitled to access to energy as a fundamental right. This can only occur if the final consumers are given top priority in the development, management, and operation of the useful energy generation process. Little progress has been made despite a great deal of well-intentioned effort, so a radical new strategy based on the following imperatives is required.

1. Rural development in general and rural energy development specifically need to be given higher priority by policymakers.
2. Rural energy development must be decentralized and local resources managed by rural people.
3. Rural energy development must be integrated with other aspects of rural development, overcoming the institutional barriers between agriculture infrastructure and education as well as in the social and political spheres.
4. The problem of non-committing policies on the government level, and different priorities in each country should be addressed (Figure 1.15).

The productivity and health of a third of humanity are diminished by a reliance on traditional fuels and technologies, with women and children suffering the most. Current methods of energy production, distribution, and use worldwide are major contributors to environmental problems including global warming and ecosystem degradation at the local, regional, and global levels. The IEC idea includes farms or dispersed living spaces where basic needs like food, energy, and water can be produced locally with little help from outside energy sources. At the IEC, energy production and consumption must be sustainable, friendly to the environment, and primarily derived from renewable sources.

Figure 1.15. *The safety standard changes for design and implementation of safety instrumented systems (SIS).*

Source: By Michael Kauffmann – Own work, CC BY 2.0 de, https://commons.wikimedia.org/w/index.php?curid=3529167.

It encompasses a variety of options for producing energy without polluting the environment, including the production of electricity from biomass and contemporary forms of wind and solar power. It should prioritize achieving ecological semi-closed systems and energetic autonomy while also ensuring socioeconomic viability and taking into account the most recent ideas in landscape and biodiversity management (Armstrong and Hamrin; TERI, 2000; Moni, 2011).

1.4. CONCEPT OF CARBON FOOTPRINT

The total amount of greenhouse gas (GHG) emissions connected to the lifecycle of these energy sources is referred to as the "carbon footprint" for renewable energy sources. Renewable energy sources, like solar, wind, hydro, and geothermal, have a certain amount of carbon footprint even though they are generally less polluting than fossil fuels because of a variety of factors related to their installation, operation, and decommissioning phases. The emissions from the extraction of raw materials, manufacturing, transportation, installation, operation, and decommissioning are all included in the carbon footprint of renewable energy sources. Even so, the carbon footprints of renewable energy sources—such as solar, wind, and hydro—are typically substantially smaller than those of fossil fuels. They offer considerable long-term reductions in GHG emissions while producing very little direct emissions while in operation. Technological developments and more effective procedures are contributing significantly to the reduction of global carbon footprints because they lower the total lifecycle emissions of these energy sources.

Here's a breakdown of how the carbon footprint of renewable energy sources can be evaluated:

1. Lifecycle Stages

   i. **Raw Material Extraction**: The process of mining and processing raw materials needed for renewable energy systems, like metals for solar panels or turbines, involves energy use and emissions.

   ii. **Manufacturing**: The production of renewable energy components (e.g., solar panels, wind turbines) requires energy. This stage can involve significant emissions if the energy used comes from fossil fuels.

   iii. **Transportation and Installation**: Transporting materials and installing renewable energy systems involves emissions, especially if heavy machinery is used.

   iv. **Operation and Maintenance**: While renewable energy systems produce minimal direct emissions during operation, there are still emissions associated with maintenance and operational energy use.

   v. **Decommissioning**: At the end of their lifecycle, renewable energy systems need to be dismantled and recycled or disposed of, which can also produce emissions.

2. Comparative Emissions

The carbon footprint of renewable energy sources is typically much lower than that of fossil fuels. For instance:

   i. **Solar Power**: The carbon footprint of solar energy comes mainly from the production of photovoltaic panels. However, once installed, solar panels generate electricity with no direct emissions. Over their lifetime, they typically result in significantly lower emissions compared to fossil fuels.

   ii. **Wind Power**: Wind turbines have emissions related to their manufacturing and transportation. However, once operational, they produce electricity without direct emissions, resulting in a low overall carbon footprint over their lifetime.

iii. **Hydro Power**: Hydropower plants can have emissions related to construction and the impacts of damming rivers, but their operational emissions are very low. The carbon footprint can vary based on the scale and design of the hydro facility.

iv. **Geothermal Power**: Geothermal plants have emissions associated with drilling and plant construction. However, the direct emissions from geothermal energy production are low compared to fossil fuels.

3. Context and Trade-offs

i. **Energy Payback Time**: This is the time it takes for a renewable energy system to offset the emissions produced during its lifecycle. Renewable energy systems generally have short energy payback times, meaning they generate clean energy faster than it takes to produce and install them.

ii. **Land Use and Ecological Impact**: While the direct carbon footprint of renewable energy sources is low, considerations like land use, water use (especially for hydro), and impact on local ecosystems are important for a comprehensive environmental assessment.

iii. **Technological Advances**: Improvements in technology and manufacturing processes continue to reduce the carbon footprint of renewable energy systems. For example, advances in materials and more efficient manufacturing techniques can lower emissions associated with production.

While renewable energy sources do have a carbon footprint, it is substantially lower than that of fossil fuels. Evaluating their environmental impact requires considering the full lifecycle emissions and comparing them to the benefits of reduced operational emissions and the long-term reduction in reliance on carbon-intensive energy sources.

1.4.1. Measurement, Significance, and Impact on Climate Change

An important consideration when assessing renewable energy sources' environmental benefits and their potential to combat climate change is their carbon footprint. Evaluating the entire lifecycle of renewable energy technologies' GHG emissions is part of the measurement process. This covers the emissions from the extraction of raw materials, production, shipping, installation, use, and decommissioning. For example, although a large amount of energy and resources are needed to produce solar panels and wind turbines, the emissions produced during their operation are minimal.

The importance of renewable energy comes from its capacity to produce electricity while emitting little or no direct carbon dioxide during operation. Renewable energy sources provide a cleaner alternative to fossil fuels, which when burned release significant amounts of CO_2 and other greenhouse gases into the atmosphere. Reducing emissions is essential to reducing climate change's effects and enhancing air quality.

There is a significant impact on climate change. We can greatly reduce GHG emissions by switching to renewable energy sources from fossil fuels, supporting international efforts to slow down global warming and lessen the risks associated with it. A more sustainable future, a reduction in the carbon footprint of energy production, and the achievement of climate goals all depend on this shift.

Long-term changes in temperature and weather patterns are referred to as climate change. These changes may be caused by significant volcanic eruptions or variations in the sun's activity. However, human activity has been the primary cause of climate change since the 1800s, mostly as a result of the burning of fossil fuels like coal, oil, and gas.

Fuel combustion releases GHG into the atmosphere, which encircles the planet like a blanket and traps solar heat, causing temperatures to rise. Methane and carbon dioxide are the two primary GHG responsible for climate change. These result, for instance, from burning coal to heat a building or gasoline to operate a vehicle. Carbon dioxide can also be released through land clearing and forest destruction. Two of the main sources of methane emissions are oil and gas operations and agriculture. Land use, buildings, transportation, energy, and agriculture are some of the primary industries producing greenhouse gases.

Scientists studying climate change have demonstrated that, for the past 200 years, almost all global warming has been caused by humans. Greenhouse gases produced by human activities such as the ones listed above are warming the planet more quickly than it has warmed in the previous two millennia. The Earth's surface is currently 11.1□ warmer on average than it was during the Industrial Revolution in the late 1800s and warmer than it has been in the previous 100,000 years. The last four decades have all been warmer than any other decade since 1850, with the most recent decade being the warmest on record, from 2011 to 2020. Many believe that the primary effect of climate change is increased temperatures. However, the story doesn't end with a rise in temperature. Since everything on Earth is interconnected, changes in one area can have an impact on changes in all other areas because the Earth is a system. Intense droughts, water scarcity, severe fires, rising sea levels, flooding, melting polar ice, catastrophic storms, and declining biodiversity are just a few of the effects of climate change that exist today.

Climate change can have an impact on our jobs, housing, safety, food production, and general health. Some of us, like those who reside in small island nations and other developing nations, are already more susceptible to the effects of climate change. Prolonged droughts are increasing the risk of famine, and conditions like sea-level rise and saltwater intrusion have progressed to the point where entire communities have had to relocate. There will likely be a rise in the number of people displaced in the future due to weather-related incidents (Armstrong and Hamrin; TERI, 2000; Moni, 2011).

1.4.2. How Is Climate Change Measured Over Time?

A place's overall weather patterns over a long period are referred to as its climate. For instance, in the United States, South Florida experiences year-round tropical weather, but Maine experiences frigid, snowy winters. The current weather and climate on Earth are tracked by satellites in orbit, distant weather stations, and ocean buoys; however, the information that allows scientists to extend the Earth's climatic records back millions of years is known as paleoclimatology, and it comes from natural sources such as ice cores, tree rings, corals, and sediments from lakes and oceans. These archives offer a thorough examination of the long-term alterations to the cryosphere (frozen water systems), land surfaces, oceans, and atmosphere of the Earth. Scientists use these data to feed complex climate models, which have remarkable accuracy in projecting future trends in climate. Though the terms "global warming" and "climate change" are frequently used synonymously, it's important to remember that global warming—the recent increase in the average global temperature close to the earth's surface—is only one component of climate change.

1.4.3. Causes of Climate Change

The Earth's climate system operates on straightforward principles. The planet cools when solar energy is reflected off the surface and back into space (mostly by clouds and ice) or when energy is released by the earth's atmosphere. The planet warms when it absorbs solar radiation or when greenhouse gases in the atmosphere stop heat emitted by the earth from radiating into space (Figure 1.16). Numerous elements, both man-made and natural, can affect the Earth's climate system.

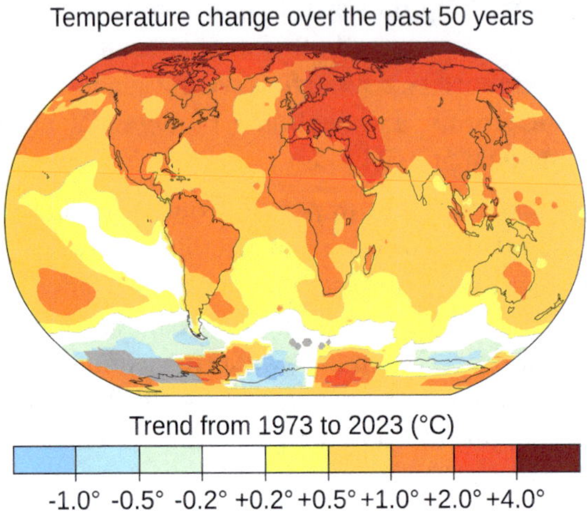

Figure 1.16. *Climate change.*

Source: By NASA's Scientific Visualization Studio, Key, and Title by uploader (Eric Fisk) – https://data.giss.nasa.gov/gistemp/maps/index_v4.html, CC BY-SA 4.0, https://commons.wikimedia.org/w/index.php?curid=113131169.

1.4.3.1. Natural Causes of Climate Change

Long before there were any humans, the earth experienced periods of warming and cooling. The intensity of the sun, volcanic eruptions, and variations in the concentrations of naturally occurring greenhouse gases are some of the factors that can cause climate change. However, historical data show that the current global warming—especially that which has happened since the middle of the 20th century—is occurring far more quickly than it has in the past, and natural causes are unable to account for it. NASA states that although these natural causes are still at work today, their impact is either too little or too sluggish to account for the recent and extreme warming."

1.4.3.2. Anthropogenic Causes of Climate Change

The primary cause of the earth's current rapid climate change is human activity, specifically the emissions of GHGs that result from it. The planet needs greenhouse gases to stay warm enough for human habitation. But in recent decades, the concentration of these gases in our atmosphere has increased dramatically. As stated by the U.S. According to the Environmental Protection Agency, our present levels of methane, nitrous oxide, and carbon dioxide "are unprecedented compared with the past 800,000 years." In fact, since preindustrial times, the amount of carbon dioxide in the atmosphere—the planet's primary contributor to climate change—has increased by 46%.

The main cause of emissions produced by humans is the burning of fossil fuels like coal, oil, and gas for transportation, heat, and electricity. Deforestation releases sequestered (or stored) carbon into the atmosphere, making it a second significant source. An average of 8.11 billion metric tons of carbon dioxide are released annually by logging, clearcutting, fires, and other types of forest degradation; this amounts to more than 20% of all carbon dioxide emissions worldwide. The use of fertilizer, which is a major source of nitrous oxide emissions, the raising of livestock (cattle, buffalo, sheep, and goats are major methane emitters), and some industrial processes that release fluorinated gases are other human activities that contribute to air pollution. The reflectivity of the earth's surface can be altered by human activities such as road construction and agriculture, which can result in localized warming or cooling.

Even though photosynthesis and other processes allow our planet's forests and oceans to absorb greenhouse gases from the atmosphere, these natural carbon sinks are unable to keep up with our planet's rising emissions. The accumulation of greenhouse gases results in warming the planet at an alarmingly rapid rate. The average temperature of the earth is thought to have increased by roughly 1°F during the 20th century. If that doesn't seem like much, keep in mind that average temperatures were only 5 to 9 degrees lower than they are today when the last ice age ended and more than 3000 feet of ice covered the northeastern United States (Armstrong and Hamrin; TERI, 2000; Moni, 2011).

1.4.4. The Effects of Global Climate Change

According to the World Economic Forum's Global Risks Report 2021, the failure to mitigate and adapt to climate change is "the most impactful" risk facing communities worldwide—ahead of even weapons of mass destruction and water crises. Blame its cascading effects: As climate change transforms global ecosystems, it affects everything from the places we live to the water we drink to the air we breathe.

And though climate change affects everyone in some way, it's indisputable that its most negative impacts are borne disproportionately by certain groups: women, children, people of color, Indigenous communities, and the economically marginalized. Climate is a human rights issue.

1.4.4.1. Extreme Weather

Wet areas get wetted and dry areas get drier as a result of the earth's atmosphere collecting, holding onto, and releasing more water as it warms. Storms, floods, heat waves, and droughts are just a few of the many natural disasters that get worse and occur more frequently when temperatures rise. These occurrences can have disastrous and expensive effects, putting people's access to clean drinking water in jeopardy, igniting uncontrollably spreading wildfires, causing property damage, spilling dangerous materials, contaminating the air, and possibly even killing people. Climate change and air pollution are closely related, with one aggravating the other.

In addition to the increase in smog and soot levels that accompanies rising temperatures, higher temperatures also lead to an increase in allergenic pollutants, such as airborne mold (resulting from damp conditions due to extreme weather and increased flooding) and pollen (due to longer and more intense pollen seasons).

1.4.4.2. Health Risks

Between 2030 and 2050, "climate change is expected to cause approximately 250,000 additional deaths per year," according to the World Health Organization. The number of deaths and illnesses from heat stress, heatstroke, cardiovascular disease, and kidney disease increases along with global temperatures. Additionally, respiratory health deteriorates with increased air pollution, especially for the 300 million people who suffer from asthma globally. Additionally, hay fever and allergy sufferers are subjected to increased levels of mold and airborne pollen. Severe storms, flooding, and other extreme weather conditions can cause harm, contaminate drinking water, and cause storm damage that could jeopardize essential infrastructure or force a community to relocate. Historical models indicate that the probability of experiencing a disaster has increased by 60% in the last 40 years—with weather- and climate-related events accounting for the majority of these increases in displacement. It is important to remember that there are health risks associated with relocation, including increased urbanization, trauma, social unrest, a lack of clean water, and the spread of infectious diseases. Insect-borne illnesses like West Nile virus, Lyme disease, and dengue fever benefit greatly from a warmer, wetter globe.

1.4.4.3. Rising Seas

The Arctic is heating up twice as fast as any other place on the planet. As its ice sheets melt into the seas, our oceans are on track to rise by anywhere from 0.95 to 3.61 feet by the end of this century, threatening coastal ecosystems and low-lying areas. Island nations face particular risk, as do some of the world's largest cities, including New York City, Miami, Mumbai in India, and Sydney in Australia.

1.4.4.4. Warmer, More Acidic Oceans

The oceans now contain 30% more acidity than they did before the Industrial Revolution because they absorb between 25% and 33% of the emissions from fossil fuels. The underwater life is seriously threatened by this acidification, especially the calcified shells and skeletons of creatures like coral, clams, and oysters. The effects on shellfish, fish, birds, and mammals that rely on shellfish for food, can be disastrous. This effect also extends to human populations in coastal communities where fishing and seafood production are the main sources of income, destroying livelihoods and creating opportunities for economic ruin. In addition to changing the distribution and population of marine life, rising ocean temperatures are also causing coral bleaching events, which have the potential to wipe out entire reefs—ecosystems that are home to over 25% of all marine life.

1.4.4.5. Imperiled Ecosystems

Wildlife is under more pressure than ever to quickly adapt to changing habitats due to climate change. Numerous species are changing their seasonal behaviors, rearranging traditional migration patterns, and moving toward colder climates and higher altitudes. These changes have the power to drastically alter entire ecosystems as well as the complex webs of life that rely on them. Consequently, a 2020 study found that by 2070, one-third of all animal and plant species may go extinct. Mammals, fish, birds, reptiles, and other vertebrate species are going extinct far more quickly than they should, according to a different study. This phenomenon has been connected to threats such as deforestation, pollution, and climate change. Experts on biodiversity and climate change released a joint report at the beginning of 2021 that highlighted these connections and called for concurrent action on both fronts. Conversely, longer summers and milder winters have made it possible for certain species to flourish, such as insects that decimate trees, putting entire forests in jeopardy (Armstrong and Hamrin; TERI, 2000; Moni, 2011).

1.4.5. Renewable Energy and Climate Change Mitigation

Because renewable energy provides a sustainable substitute for fossil fuels, which are the main source of greenhouse gas emissions, it is essential for mitigating the effects of climate change. The main advantage of renewable energy sources, which include hydro, geothermal, wind, and solar, is their capacity to produce electricity with negligible or no direct carbon emissions. When burning, fossil fuels release large amounts of CO_2 and other GHGs; in contrast, renewable energy sources use natural processes that have very little effect on the environment while in use.

We greatly reduce greenhouse gas emissions by switching to renewable energy sources from carbon-intensive ones, which is essential for reducing global warming. This shift serves to slow down the rate of temperature rise as well as the frequency and intensity of climate-related effects like extreme weather, sea level rise, and ecosystem disruption. Furthermore, by encouraging a transition to greener energy systems and stimulating innovation in energy technologies, the integration of renewable energy contributes to the achievement of long-term climate goals, such as those outlined in the Paris Agreement.

Furthermore, by diversifying energy sources and lowering reliance on imported fuels, which are sometimes vulnerable to unstable markets and supply disruptions, renewables help to mitigate the effects of climate change. This supports sustainable economic growth in addition to improving energy security. In conclusion, renewable energy provides a cleaner, more sustainable way to meet our energy needs while preserving the environment, making it a crucial part of the global strategy to combat climate change.

1.4.6. How Renewable Energy Reduces Carbon Emissions

Renewable energy reduces carbon emissions through several key mechanisms:

1. **Zero Emissions During Operation**: Unlike fossil fuels, renewable energy sources like solar, wind, hydro, and geothermal generate electricity without emitting carbon dioxide (CO_2) or other GHGs during operation. This direct reduction in emissions is a primary way renewables help mitigate climate change.
2. **Displacing Fossil Fuels**: By providing an alternative to coal, oil, and natural gas, renewable energy sources reduce the need for burning fossil fuels, which are major contributors to CO_2 emissions. When renewables replace fossil-fuel-based power plants in the energy mix, overall carbon emissions from the energy sector decrease.
3. **Improving Energy Efficiency**: Many renewable energy technologies, particularly wind and solar, have become increasingly efficient, meaning they can produce more energy with fewer resources and less environmental impact. This enhanced efficiency contributes to a lower carbon footprint.
4. **Reducing Energy Storage and Transmission Losses**: Advances in energy storage and smart grid technologies enable better integration of renewables, reducing the reliance on backup fossil fuel plants and minimizing energy losses. Efficient energy storage solutions help ensure a stable energy supply, further reducing the need for carbon-intensive backup power sources.

5. **Supporting Sustainable Practices**: The deployment of renewable energy often goes hand-in-hand with other sustainable practices, such as energy conservation, efficiency improvements, and reduced deforestation. These practices collectively contribute to lower overall carbon emissions.

Renewable energy sources reduce carbon emissions by generating power without direct emissions, displacing fossil fuels, improving energy efficiency, and supporting broader sustainability practices. This comprehensive impact is essential for addressing climate change and achieving long-term environmental goals (Armstrong and Hamrin; TERI, 2000; Moni, 2011).

REFERENCES

1. Armstrong, A. J., & Hamrin, J. (n.d.). *The Renewable Energy Policy Manual*. Organization of American States.
2. IbaUsa. (2009). *India Energy Policy, Laws, and Regulations Handbook: Vol. I. Strategic Information and Renewable Energy*. International the Science of Business Publication.
3. Madke, N. S., & Bhise, V. B. (2002). *Economics of Biogas Technology*. The Associated Publishers.
4. Moni, D. R. (2011). *Energy Sources and Policies in India*. New Country Publications.
5. Mumoki, F. (2006). The effects of deforestation on our environment today. *Panorama*. TakingITGlobal. Retrieved from: https://www.takingitglobal.org (accessed on 9 November 2024).
6. Sinha, P. C. (2009). *Handbook of Alternate Energy*. SBS Publishers & Distributors Pvt. Ltd.
7. Stead, J. G., & Stead, W. E. (2009). *Management for a Small Planet*. M.E. Sharpe.
8. TERI. (2000). *TERI Energy Data Directory and Yearbook*. The Energy Research Institute.
9. Walton, A. L., & Hall, D. O. (1990). Solar power. *Contemporary Policy Issues*, *8*(240–254).
10. Williams, R. H., & Larson, E. D. (1994). Advanced gasification-based biomass power generation. In T. B. Johansson et al. (Eds.), *Renewable Energy Sources for Fuels and Electricity* (pp. 109–120). Island Press.

Chapter 2

Solar Energy

Contents

INTRODUCTION

An alternative energy source called solar energy works by collecting the sun's radiant light energy and turning it into electrical current. With significant advancements in solar energy harnessing and utilization since the mid-1900s, homes, and businesses can now use this renewable energy source instead of relying on more traditional power generation methods. The use of solar energy is still expanding, as is the creation of more affordable methods for storing and capturing solar energy for later use. Currently, using a system with multiple solar panels and storage batteries is the most popular way to capture solar energy. The panels gather radiant light, and the batteries store the energy they have collected. Energy can be used in real-time to run a variety of machines and household appliances in addition to being stored. The extra is kept in reserve for use at night or in other circumstances where there isn't easy access to radiant light.

The operation of appliances like stoves, refrigerators, computers, and television sets as well as entertainment devices can now be sustained by batteries in a home powered by solar energy. In addition, solar-powered houses can use their stored energy to run water heaters or to heat and cool their houses. These days, some houses have hybrid power systems that combine the use of solar energy with electricity from a conventional power grid. Although the house is not entirely reliant on solar energy, this kind of system can reduce utility costs and offer a reliable backup if a portion of the nearby power grid fails. Healthcare facilities are starting to consider solar energy as a backup power source in an emergency, just like homes. This would enable a hospital to keep running even if the backup generator system could not keep up with the current power demand and the power failed for any reason.

These days, solar energy systems are much less expensive to build and install than they were in the last decades of the 20th century. The newer generation of solar energy systems are not only more cost-effective than their predecessors, but they are also significantly more efficient, making them more desirable to both home and business owners.

2.1. CONCEPT OF SOLAR ENERGY

If nothing else, we have been researching solar energy extensively in recent years to find a more environmentally friendly means of obtaining energy. For those who are unfamiliar with solar energy, it is essentially the energy that we extract from the sun and transform into useful energy that can be used for heating, cooling, or producing electricity.

To put it simply, solar energy is energy that originates from the sun. The sun is a potent energy source. There wouldn't have been any life on Earth without it. Many home and business owners currently use this clean, renewable, and freely available energy source for a range of uses (Figure 2.1).

Figure 2.1. *A worker cleans the panels in a solar power station.*

Source: By Lucas Braun – Own work, CC BY-SA 3.0, https://commons.wikimedia.org/w/index.php?curid=12261041.

The best and most accessible energy source on Earth is solar energy. It is widely used in cooking, water heating, power generation, distillation, cold storage, and other processes. The amount of nonrenewable energy resources consumed can be significantly decreased by making effective use of solar energy.

2.1.1. How Does Solar Energy Work?

Solar energy is all about converting the energy from the sun into energy that we can use. There are three major ways that we can do this (Figure 2.2).

Figure 2.2. *Solar panels and how they work to create eco-friendly energy in your home.*

Source: By PeterVauhn – Own work, CC BY-SA 4.0, https://commons.wikimedia.org/w/index.php?curid=80406131.

2.1.1.1. Photovoltaic Processes (PV)

1. PV processes happen with PV devices, which take the sunlight and make it into electricity.
2. PV processes use materials that are known for naturally changing sunlight into electricity, which is known in the science world as semiconductors.
3. When the sunlight hits these semiconductors, they produce electrons, which are pushed through the device and make a current.
4. PV devices are still in development at this point in time; some rays of light, including UV and infrared, cannot be utilized by PV devices. This can be problematic when the sun is not shining brightly and these are the main rays that are being produced.

2.1.1.2. Solar Heating and Cooling (SHC)

1. Solar heating and cooling (SHC) is all about thermal energy.
2. The energy that is created from the sun is switched into energy that can be used to either heat or cool the devices that are being powered.

3. At this point, some processes can help in almost any context that you can imagine, from commercial buildings to homes.
4. The main reason that SHC devices are necessary is that we need to reduce our dependence on natural gases and other fossil fuels.
5. Technology in the SHC sector is always changing and adapting, and advances are being made in this area of science every single day (Figure 2.3).

Figure 2.3. *Synthesis of a solar system.*

Source: By Gilabrand at English Wikipedia, CC BY 2.5, https://commons.wikimedia.org/w/index.php?curid=7954929.

2.1.1.3. Concentrating Solar Power (CSP)

1. Concentrating solar power, usually referred to as CSP, is one of the newest technologies that are out there, even though the concept is not as new as many think it is. People have used the concept behind it for a very long time, which is why it's shocking that it's taken so long for us to actually harness it on a larger scale (Figure 2.4).
2. Mirrors are the main focus of CSP plants; they concentrate the energy from the sun to cause everything to work effectively. Without the mirrors, CSP systems would not work as effectively as some of the other systems that are currently out on the market today.

3. The sun's energy then makes steam engines and turbines go so that the electricity can be produced and stored. This makes it different than the other types because the storage system is a lot more advanced than what you see from the other types of solar energy that are currently being used.
4. Currently, there are just over 1500 different CSP plants located throughout the entire United States; there are over 400 more to be planned within the next 5 years. By the 2020s, the plan is to have over 3000 CSPs working throughout the country, with over 5000 throughout the entire world (Scheer, 2002; Mills, 2004).

Figure 2.4. *Mohammed bin Rashid Al Maktoum Solar Park in Dubai is home to the world's largest solar power project with a capacity of 700 megawatts and to be built for Dh14.2 billion.*

Source: By Contains modified Copernicus Sentinel data 2020, Attribution, https://commons.wikimedia.org/w/index.php?curid=92808814.

2.1.2. How Solar Energy Is Converted to Electricity?

Installing photovoltaic (PV) or solar cells is the first step in converting solar energy to electricity. PV simply means electricity and light (Figure 2.5). These solar cells capture solar radiation and transform it into electrical power. These solar cells are constructed from materials that exhibit the PV effect, which means that when sunlight strikes the cell, photons of light inside the cell spook the electrons, causing them to begin flowing and ultimately producing electricity.

When looking to buy solar panels, it would be beneficial to know the options available in the marketplace. Here is an outline of the main ones:

1. **Polycrystalline** – This utilizes multicrystalline silicon.
2. **Monocrystalline** – This is ideal for small spaces.

3. **Thin Film** – Typically bigger in size and a lot more efficient during the day.

Figure 2.5. *Home-based solar system: Renewable energy solutions for residential users.*

Source: By Andrewglaser at English Wikipedia, CC BY-SA 3.0, https://commons.wikimedia.org/w/index.php?curid=30358256.

The composition of the silicon substrate used to create solar cells and, eventually, solar panels accounts for the difference in the material used to create monocrystalline and polycrystalline solar panels. Polycrystalline refers to many crystals, whereas monocrystalline refers to a single crystal. The reason why monocrystalline cells are typically 10–15% more efficient than polycrystalline crystals is because larger crystal sizes translate into more efficient solar cells (Scheer, 2002; Mills, 2004).

2.1.3. Solar Panel

An apparatus called a solar panel is designed to collect and store solar energy for later use. Solar thermal collectors and photovoltaic modules are the two primary categories of solar panels. The purpose of photovoltaic modules is to transform solar energy into electrical energy for use in devices like computers, lights, and automobiles. Water for showers or water that can be forced through a structure to heat it is frequently heated using solar thermal collectors, which are designed to directly absorb solar energy as heat energy. The photovoltaic module, the earliest kind of solar panel, is essentially an assembly of several solar cells.

A solar array, also called a photovoltaic array, is made up of several solar panels (Figure 2.6). For convenience of use and economies of scale, solar cells are integrated into a single solar panel. They become easier to transport and install when combined into manageable panels, and multiple solar cells can share the electronics needed to process them. The most popular kind of solar panel uses thin films composed of either silicon or cadmium telluride, or crystalline silicon in the form of wafers.

A portion of the light energy that strikes the solar panel is absorbed by the semiconductor that makes it up, releasing electrons. After being driven in a single direction, these electrons produce a current that can be captured, transformed, and utilized to power various devices. Although it's a very simple idea, a fascinating world of invention and research is now being conducted to fully realize the potential of solar panels.

Figure 2.6. *It is our belief that the sun can, and will, power our lives, and solar photovoltaic (PV) panels will be a major part of that transition.*

Source: By Unknown author – Department of Energy Retrieved Aug. 17, 2005., Public Domain, https://commons.wikimedia.org/w/index.php?curid=461648.

Getting the most energy out of the smallest possible area is one of the most pressing issues with solar panels. Installing solar panels instead of using electricity from the grid must be financially and environmentally advantageous because they are costly and require space. The efficiency range for consumer solar panels is currently between 5% and 18%, and over the last 10 years, this range has grown considerably. There is a thriving industry devoted to creating cutting-edge solar panels with even higher absorption rates, and it appears probable that in the not too distant future, reasonably priced consumer solar panels with absorption rates of 20–30% will be accessible.

More ambitious plans are also in the works to increase efficiency significantly; many theoretical designs aim for as much as 50% efficiency. This has ramifications for many different applications, such as powering houses, cars, and spacecraft. To mount solar panels on objects like windows and essentially erase their footprint, another advancement in solar panel design aims to make the panels themselves exceedingly thin, transparent, and flexible.

The solar thermal collector, the alternative kind of solar panel, has been in use in one form or another for hundreds of years. Essentially, the concept involves submerging a body of water in a container and maximizing its absorption of light energy—this can be achieved, for example, by painting the container black. After that, the water is heated and ready for use for a variety of uses. Many use forced solar heating, which involves pumping hot water through pipes in the walls or floors, where it radiates heat into the house and heats the air within (Scheer, 2002; Mills, 2004).

2.1.4. How Do Solar Panels Work?

The same electronics that power chemical batteries or conventional electrical outlets also power solar panels, which can be found on calculators and space stations alike. It all comes down to the free passage of electrons through a circuit in solar panels.

A brief trip back to high school chemistry class might be helpful to grasp how these panels produce electricity (Figure 2.7). Pure silicon, which contributed to the computer revolution, is the fundamental component of solar panels. Removed of all impurities, silicon provides an excellent neutral medium for electron transfer. Moreover, silicon has certain atomic-level characteristics that increase its appeal for use in solar panel manufacturing.

Although silicon atoms only carry four electrons in their native state, they have space for eight electrons in their outer bands. This indicates that four more electrons can fit in. The four electrons of the other silicon atom are shared by the two that come into contact with it. Although a strong bond is formed, the atoms' needs are met by the eight electrons, so there is no positive or negative charge. Years can pass while silicon atoms combine to form a sizable chunk of pure silicon. The panels' plates are formed from this material. This is where science comes into play. Since pure silicon has no positive or negative charge, two silicon plates together cannot produce electricity in solar panels. Silicon is combined with other elements that carry positive or negative charges to create solar panels.

For example, phosphorus can give other atoms five electrons. A chemical combination of silicon and phosphorus yields a stable eight electron system with one free electron tagging along. It is bonded to the other phosphorus atoms and cannot escape, but silicon does not require it. As a result, it is believed that this new silicon/phosphorus plate is negatively charged.

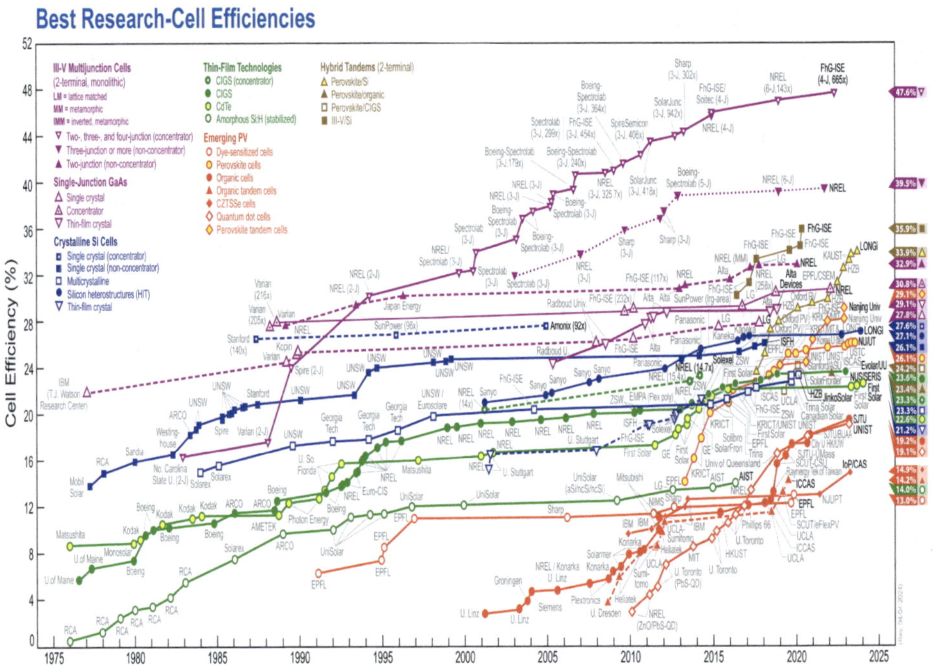

Figure 2.7. *It is amazing that we can convert sunlight into energy and electricity through the use of solar panels. Read more and discover how solar panels work.*

Source: By Nikos Kopidakis – National Renewable Energy Laboratory (NREL), Golden, CO, Public Domain, https://commons.wikimedia.org/w/index.php?curid=126894974.

There must also be a positive charge for electricity to flow. This is accomplished by mixing silicon with an element that can only offer three electrons, like boron. One electron is still available on a silicon/boron plate. This indicates a positive charge on the plate. Conductive wires are positioned between the two plates that are sandwiched together in the panels. Now that the two plates are installed, let's discuss the "solar" side of solar panels.

Although a variety of energy particles are released by natural sunlight, the one in which we are most interested is known as a photon. In essence, a photon functions like a moving hammer. Photons attack the silicon and phosphorus atoms in solar cells when their negative plates are oriented correctly with respect to the sun. Eventually, the ninth electron—which is inherently free—is removed from the outer ring.

The positive silicon/boron plate pulls this electron into the open spot on its own outer band, so it does not stay free for very long. Electricity is produced as more electrons are broken off by the photons in the sun. One solar cell doesn't produce a lot of electricity, but when all the conductive wires pull the free electrons from the plates, enough electricity is produced to run other electronics or low-amperage motors. The process starts over on the negative plate when any electrons that are not utilized or lost to the atmosphere are brought back.

The fact that solar panels produce relatively little electricity concerning their size is one of the primary issues with using them. A solar-powered car would need several thousand solar cells, whereas a calculator might only need one. The efficiency of the panels can decrease by 50% if the angle is altered, even slightly.

Chemical batteries can store a portion of the power generated by solar panels, but in most cases, there isn't much extra power to begin with. Photons are produced by sunlight, but it also emits more harmful UV and infrared waves that eventually lead to physical degradation of the panels. Additionally, the panels need to be exposed to harsh weather conditions, which can have a negative impact on their efficiency.

A lot of sources also call solar panels PV cells, alluding to the role that light (photos) plays in producing electrical voltage. The task facing scientists in the future will be to develop more powerful panels that can generate excess energy during periods when sunlight is not available, while also being small enough for practical applications.

The solar panel ought to be installed in a clear space free from obstructions like trees or other fixtures. Usually, a roof is your best option. After that, the inverter is used to wire it into the structure. An apparatus that changes alternating current (AC) into direct current (DC) is called an inverter. Therefore, in this instance, the energy produced by the solar panels is the AC. AC is produced from this solar energy. The energy is converted from direct current to AC so that it can be used by different household appliances in the same manner as regular electricity powers your electronics.

2.4.1.1. Considerations to Keep in Mind Before Installing Solar Panels

Make sure solar power is appropriate for you and your house before installing solar panels. Next, confirm that there is enough sunlight in your area. The amount of sun that a region receives has a significant impact on how suitable solar energy is. Purchasing solar panels is not the best course of action if your area does not receive enough sunlight. Once you've established that your location receives enough sunlight, confirm that there is room for installation. We now know that solar panels are primarily installed on building roofs. Although commendable, this is not the sole choice. Ground mounts would be perfect in your backyard if you have any empty space there. For people whose roofs are heavily shaded or are not structurally necessary for solar panels, the backyard option is perfect. To keep out of trouble with the local authorities, it is also essential to be aware of the laws in your area regarding the installation of solar panels. Your local solar energy consultant has this information (Scheer, 2002; Mills, 2004).

2.1.5. Thermodynamics of Solar Energy

Energy, as we all know, is a basic necessity for life. The most abundant source of energy on Earth is sunlight. Every living thing is reliant on solar energy. Nevertheless, 99% of incident solar energy is lost to the environment as heat and is never used by living things.

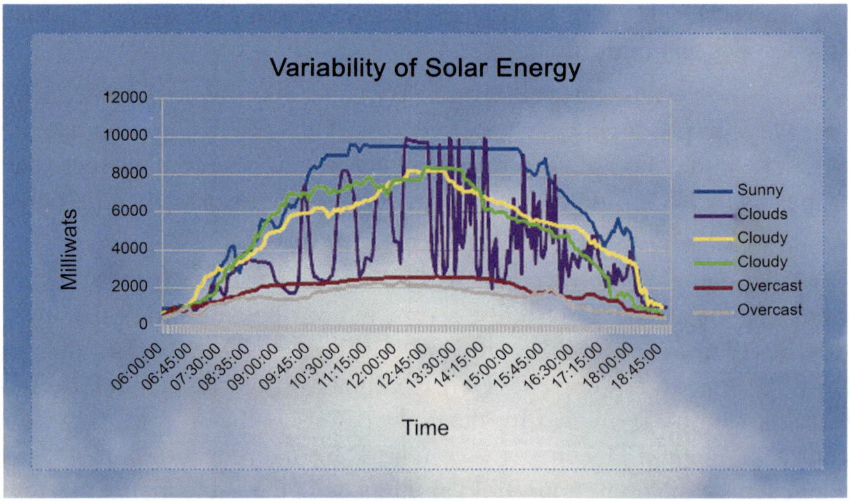

Figure 2.8. *Variability.*

Source: By CharlesMJames – Own work, CC BY-SA 4.0, https://commons.wikimedia.org/w/index.php?curid=102757262.

Green plants only capture around 1% of the total light energy, which is then stored as ATP chemical energy. The two fundamental laws of thermodynamics are included in the thermodynamics of solar energy. The first law, which states that matter and energy cannot be created or destroyed, is the law of conservation of matter and energy. It is possible to change matter and energy from one form to another. As a result, the total incident solar energy is equal to the energy that living things use and the energy that is released into space as heat. The law of entropy is the second law of thermodynamics (Figure 2.8).

2.1.6. Power from Solar Photovoltaic

Photovoltaic refers to the process of directly converting light energy into electrical current at the atomic level. The process known as the "photovoltaic effect" occurs when a material with certain properties is used in PV cells to absorb photons of light and release electrons. After being captured, these electrons produce electric current, which is subsequently used to generate electricity. The semiconductor materials used to make solar cells, or basic PV cells, create an electric field with a positive charge on one side and a negative charge on the other. When light energy collides with a semiconductor material's surface, electrons are liberated from their atoms. To capture the released electrons as an electric current, electrical conductors are connected to the positive and negative sides of an electric circuit, that is, electricity (Figure 2.9).

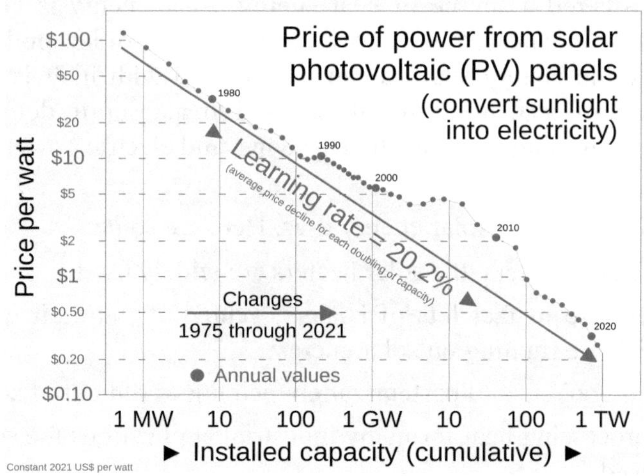

Figure 2.9. *Solar photovoltaic power tops the globally installed power rankings.*

Source: By RCraig09 – Own work, CC BY-SA 4.0, https://commons.wikimedia.org/w/index.php?curid=138292064.

Numerous solar cells that are electrically connected to one another make up a PV module. Modules can be tailored to operate as a source of electricity at a specific voltage. The amount of light that strikes the solar PVmodule determines how much power it can produce. Many modules are connected by wire to produce power in large quantities because increasing the surface area available for light strikes also increases power production.

2.1.7. Nature of Solar Radiation

The nature of solar radiation can be described by its wavelength. Solar radiation occupies the region from approximately 0.1 to 4.0 micrometers in an electromagnetic spectrum. It consists of visible light, ultraviolet light, near infrared and far infrared radiation. The 44 part of solar radiation falls in the visible light zone of the electromagnetic spectrum. The form of electromagnetic radiation that can be seen by our eyes makes the visible light. Its wavelength falls in the range of0.40 to 0.71 micrometers (μm). About 7 of the solar radiation falls in UV light. Its wavelength falls in the range of 0.1 to 0.4 micrometers. About 48 of the solar radiation fall in the range of wavelength between 0.71 to 4.0 micrometers. It constitutes the near (0.71 to 1.5 micrometers) infrared light and far infrared light (1.5 to 4.0 micrometers).

2.1.8. Uses of Solar Energy

Once only considered a source of heat energy, solar energy is now used to power numerous appliances. People nowadays are aware of its applications and how non-renewable energy sources can be replaced with it. It is even more efficient in terms of cost. It is typically utilized in spacecraft, domestic water heating systems, building space heating systems, and electric power for remote installations.

There are in fact many solar energy uses. Here are some:

1. As water heaters, the solar heaters are widely used.
2. There are in fact lots of heaters, ventilators, and air conditioners which are running on solar energy.
3. Solar cookers and pasteurization techniques run out of solar energy.
4. In processing heat for many industrial applications the solar energy is used.
5. Solar energy is used as fuel in water treatment which does disinfection, desalination of water.
6. Lastly the most important use in producing electricity, there are lots of photovoltaic cells used in running electric power used for lighting, running fans and other domestic uses (Figure 2.10).

Figure 2.10. *Get the house or any building powered by the solar panels and get the reduction in electricity bills. The power generated from the solar panel is transferred to the inverter which generates the power to run your home. You can use the same for a shed, warehouse, home, commercial building or any building.*

Source: By MdCAlmeida Villafuerte – Own work, CC BY-SA 4.0, https://commons.wikimedia.org/w/index.php?curid=44507806.

2.1.9. Environmental Impact

Despite being one of the cleanest and most renewable energy sources currently in use, solar energy does have some negative effects on the environment (Figure 2.11). Photovoltaic cells are used in solar energy to generate solar power. However, silicon is needed for the production of the photovoltaic cells, which generate some waste products as well as energy. When these materials are handled improperly, people, and the environment may be exposed to dangerous levels of contamination. Existing ecosystems may be impacted by the large amounts of land needed to install solar power plants. When solar panels are used to convert solar energy into electricity, there is no air pollution. It is widely distributed and has no mitigating effect on global warming.

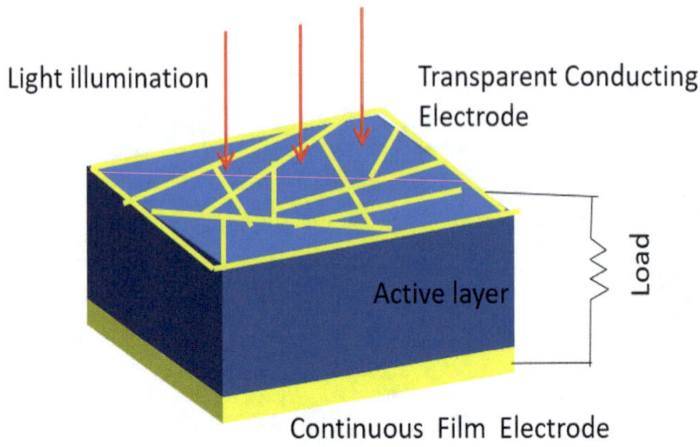

Figure 2.11. *At the individual installation level, it's rather obvious that commercial solar energy systems have a positive impact on the environment—instead of using electricity generated by fossil fuels, your business utilizes clean, renewable energy from a limitless resource (the sun).*

Source: By Photo78harry – Own work, CC BY-SA 4.0, https://commons.wikimedia.org/w/index.php?curid=71042927.

2.1.10. Advantages of Solar Energy

1. **Renewable and Sustainable:** Solar power is renewable and sustainable. The sun is going to shine till the end of this planet. It all depends on how can we make the best use of resources and use this free source of energy to power our homes and offices.
2. **Power Remote Areas:** This clean and renewable source of power has proved a boon for people living in remote areas and not necessarily connected to the national grid.
3. **No Pollution:** Solar energy is silent and produces no pollution at all. This is one of the big reasons why it is favored by many people all around the world.
4. **Surplus Electricity:** The surplus electricity that is generated can be sent back to the power grid and you can get a rebate in your monthly electricity bill.
5. **Reduce Electricity Bills:** The electricity that is generated can be used to power fans, lights, charge mobiles and much more. It is free, renewable, can be produced on demand and also helps you to bring down your monthly electricity bill.

6. **Low Maintenance:** Solar panels when installed, require little or no maintenance. They require cleaning only a few times a year.
7. **Job Creation:** The solar energy sector has opened tremendous opportunities for people. Without a doubt, it has slightly reduced the unemployment rate in a country.

2.1.11. Disadvantages of Solar Energy

1. **Initial Investment:** An initial investment is what is required if you need to harness energy from the sun. This cost can however pay itself in 10–15 years' time frame depending on the amount of energy harnessed and the money saved in buying electricity from the local grid.
2. **Available during the Daytime:** Solar energy can only be harnessed during the daytime. This simply means you can use this source of energy for half a day only.
3. **Not Effective during Cloudy or Rainy Seasons:** Cloudy and rainy seasons reduce the effectiveness of solar panels. This is the reason why solar energy is not successful in areas where it remains cloudy for most of the time of the year.
4. **Large Setup Required:** For a large business owner, a huge setup may be required to harness solar energy as its demand for energy may be more which may not be fulfilled by installing solar panels just on the rooftops.
5. **Expensive Storage:** The energy harnessed from the sun needs to be stored in the batteries to make it available for use later at home or office. The cost of these batteries may influence a buyer's decision to go for solar power (Scheer, 2002; Mills, 2004).

2.2. APPLICATIONS OF SOLAR ENERGY

The sun produces solar energy through nuclear fusion reactions. Insolation occurs at a time when the incident solar energy is not fully utilized to meet the energy demands. Solar devices are typically used to convert solar energy into thermal or electric energy. However, solar energy is only available during the hours of sunshine, and there might be a need for thermal or electrical energy during the rest of the day. Furthermore, there's a chance that the maximum solar energy availability and the demand for thermal or electric energy won't line up precisely. Cloudy days can cause solar energy availability to drop for a few days at a time, which significantly reduces the amount of thermal and electric energy produced by solar radiation. Therefore, to meet the demand for both thermal and electric loads during peak demand times, it is imperative to store the energy output (both thermal and electric) from solar devices during high insolation (in = incident, sol = solar, and ation = radiation). When there is insufficient insolation, a solar energy storage system can supply more power than the solar electric or thermal plant can produce, allowing energy production to be balanced with load demand.

Heat and light are the two primary forms of solar energy that are commonly utilized, making it a very useful and practical source. People use solar energy to heat water, charge automobiles, and produce electricity using PV cells.

Some of the major applications of solar energy are as follows:

1. Solar water heating.
2. Solar heating of buildings.
3. Solar distillation.
4. Solar pumping.
5. Solar drying of agricultural and animal products.
6. Solar furnaces.
7. Solar cooking.
8. Solar electric power generation.
9. Solar thermal power production.
10. Solar greenhouses.

2.2.1. Solar Water Heating

A flat plate metal collector that has been blackened and connected to metal tubing that faces the sun's direction makes up a solar water heating unit (Figure 2.12). Underneath the transparent glass cover of the plate collector is a layer of thermal insulation.

An insulated tank that holds hot water on cloudy days is connected to the collector's metal tubing via a pipe. The collector gathers solar radiation and uses a pump or gravity to transfer the heat to the water that is running through the tubing.

The corresponding metal tubing delivers this hot water to the storage tank. Hospitals, canteens, vacation homes, hotels, guest houses, and other residential and commercial buildings frequently use this type of water heating system.

Figure 2.12. *Solar water collectors installed.*

Source: https://upload.wikimedia.org/wikipedia/commons/3/35/Sistema_por_Hidroneumatico_%28Calentador_solar_para_alta_presi%C3%B3n%29.JPG.

2.2.2. Solar Heating of Buildings

Solar energy can be used for space heating of buildings in many ways namely:

1. Collecting the solar radiation by some element of the building itself that is, solar energy is admitted directly into the building through large South-facing windows.
2. Using separate solar collectors which may heat either water or air or storage devices that can accumulate the collected solar energy for use at night and during inclement days.

Conventional equipment like fans, ducts, air outlets, radiators, hot air registers, so forth are used to transfer heat from these collectors or storage devices to the building when it is needed. to make a building's living areas more comfortable.

The heated air or water from the collector can be transferred to a heat storage device, such as a well-insulated water tank or other heat-holding material when the building is not in use. An auxiliary heating system that runs on gas, oil, or electricity is needed as a backup system on bad days.

2.2.3. Solar-Distillation

Potable water is scarce in semi-arid and/or coastal areas. Through the process of solar distillation, salty water can be transformed into drinkable distilled water in these regions thanks to an abundance of sunshine (Figure 2.13). Using this technique, sunlight is allowed to enter a shallow, blackened basin filled with salted water through a transparent, airtight glass cover.

When solar radiation enters the brine (impure saline water) through the covers, it is absorbed, and converted to heat, and the water in the brine evaporates as a result. In the cool interior of the roof, the produced vapors condense to form pure water.

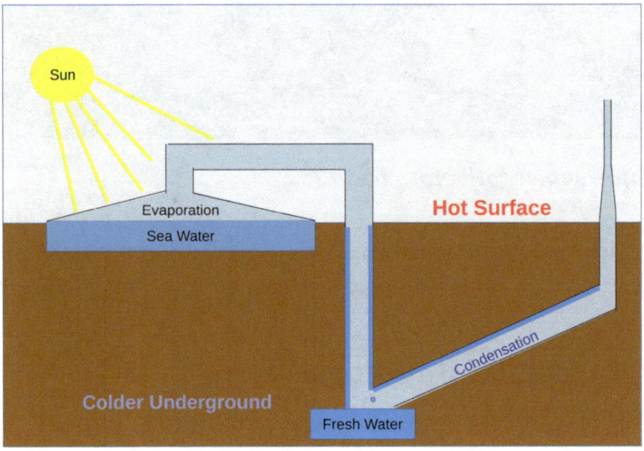

Figure 2.13. *Solar water purification system with solar heating.*

Source: By Engelbert Niehaus – LibreOffice Draw Images exported to SVG, CC BY-SA 4.0, https://en.wikipedia.org/w/index.php?curid=55237502.

Condensed water runs off the sloped roof, collects in the bottom-positioned troughs, and then fills a water storage tank to provide drinkable, deionized water to places where it is scarce, such as science labs in schools, colleges, gas stations, hospitals, and the pharmaceutical industry. When compared to other electrical energy-based processes, the cost of distilled water obtained by this system is less per liter.

2.2.3.1. Solar-Pumping

Solar pumping is the process of pumping water for irrigation using the power produced by the sun. This method works best for irrigation because the need for water pumping is highest during the hot summer months, which also happen to be the times when solar radiation is highest. The need for water pumping is comparatively lower during bad weather conditions with low solar radiation because crop transpiration losses are also low (Scheer, 2002; Mills, 2004).

2.2.4. Solar Drying of Agricultural and Animal Products

This is an old-fashioned way of drying agricultural and animal products with solar energy. A straightforward cabinet dryer that consists of a box with an inclined transparent sheet of glass covering it, painted black on the inside, and insulation at the base is used to dry agricultural products.

Ventilation holes are positioned at the base and top of the sides to allow air to pass over the drying material that is positioned on perforated trays within the cabinet. The purpose of these perforated trays or racks is to allow for controlled exposure to solar radiation.

Fruit quality is enhanced by solar drying, particularly when the fruit's sugar content rises during the drying process. Soft fruits are typically more susceptible to insect attack as their sugar content rises during the drying process. However, a fruit drier saves a significant amount of time by speeding up the drying process, which reduces the likelihood of insect attack (Figure 2.14).

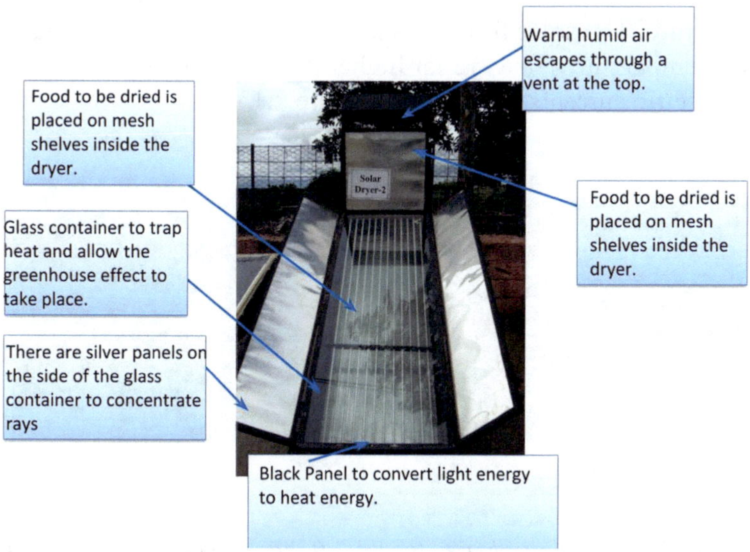

Figure 2.14. *The solar drying system utilizes solar energy to heat up air and to dry any food substance or other product loaded, which is beneficial in reducing wastage of agricultural products or any other product and helps in the preservation of the product.*

Source: https://commons.wikimedia.org/wiki/File:Solar_dryer_2.jpg.

In addition to requiring a lot of open space and manual labor for material handling, the current method of drying chilies by spreading them out on the floor makes it difficult to maintain the chilies' quality and flavor unless drying is done in a controlled environment. Furthermore, sun-dried goods are frequently ruined by birds, dust storms, and unexpected downpours. Furthermore, it is not feasible to achieve extremely low moisture content in the sun-dried chilies, according to reports.

This makes the chilies more vulnerable to fungal and bacterial attacks. Sometimes when produce is sun-dried, it gets overly dry and loses quality. A dryer powered by solar energy helps to mitigate most of these drawbacks. Potato chips, berseem, corn, and rice grains, ginger, peas, pepper, cashew nuts, timber, and veneer drying, and tobacco curing are other agricultural products that are frequently sun-dried. Animal products that are sun dried include fish and milk spray dried.

2.2.5. Solar Furnaces

A solar furnace uses many heliostats, or turn-able mirrors, arranged on a sloping surface to concentrate solar radiation onto a specimen to achieve a high temperature. The solar furnace is used to investigate the characteristics of ceramics at temperatures that are far higher than those that can be measured in labs using electric currents and flames (Figure 2.15).

Figure 2.15. *A solar furnace.*

Source: By H. Zell – Own work, CC BY-SA 3.0, https://commons.wikimedia.org/w/index.php?curid=11386331.

By shifting the material in focus, temperature can be easily controlled and heating can be carried out without any contamination. This is particularly helpful for chemical and metallurgical processes. An open specimen can be measured for a variety of properties. Future uses for solar furnaces include the air-to-fertilizer and nitric acid conversion process (Scheer, 2002; Mills, 2004).

2.2.6. Solar Cooking

For cooking, a range of fuels are used, including coal, kerosene, cooking gas, firewood, dung cakes, and agricultural wastes. Owing to the energy crisis, these fuels' supplies—wood, coal, kerosene, and cooking gas—are either running low or are too valuable to be thrown away for cooking (cow dung is better off as manure to increase soil fertility). This made the development of solar cookers and the use of solar energy for cooking necessary (Figure 2.16).

The flat plate box style solar cooker is a basic solar cooker. It is made up of an interior-blackened, well-insulated metal or wooden box. Short wavelengths of solar radiation enter the box. Reduced reradiation from the blackened interior to the outside of the box through the two glass covers reduces heat loss because higher wavelength radiation cannot pass through the glass covers.

By sealing the box tightly, convectional heat loss is reduced to a minimum. This is accomplished by placing a rubber strip between the box's upper lid and outer cover to reduce conduction-related heat loss. The area between the box's outer cover and the blackened tray is then filled with offensive materials such as sawdust, glass wool, and paddy husk.

The temperature inside the box rises as a result of solar radiation entering through the glass covers and being absorbed by the blackened surface. Pots that have been blackened on the outside are put inside the solar box.

Figure 2.16. *A solar cooker.*

Source: By Anefo – http://proxy.handle.net/10648/aa1ae048-d0b4-102d-bcf8-003048976d84, CC0, https://commons.wikimedia.org/w/index.php?curid=97889065.

The heat energy generated by the solar box's elevated temperature cooks the raw food. A plane reflector mirror can be added to such a solar cooker to increase its collector area. The temperature inside the cooker box rises by 15° to 25□ when this reflector is adjusted to reflect the sun's rays into the box.

The main benefit of using a solar cooker is that it produces no pollution, charring, or spilling of food while cooking, and the food retains all of its vitamins and natural flavors. This means that the cooked food has an extremely high nutritional value.

There is very little maintenance required for the solar cooker. The primary drawback of the solar cooker is its inability to cook food at night, on overcast days, or on short notice. Chapatis cannot be cooked in a solar cooker, and cooking takes a lot longer.

2.2.7. Solar Electric Power Generation

Photovoltaic cells can be used to directly produce electricity or energy from solar radiation. An energy conversion tool called a PV cell is used to directly convert solar photons into electrical energy. It is composed of semi-conductors that absorb photons from the sun to produce highly energetic free electrons.

An electric field causes these high-energy free electrons to flow out of the semiconductor and perform beneficial tasks. In PV cells, this electric field is typically produced by a p-n junction made of materials with various electrical characteristics. To maximize the efficiency of these cells, various fabrication techniques are used (Figure 2.17).

Cell modules are created by combining these cells in a parallel or series configuration. These modules have many unique features, including high dependability, zero fuel costs, low maintenance costs, long lifespan, portability, modularity, and pollution-free operation.

Irrigation pumps, automated weather stations, navigational signals, railroad crossing alerts, emergency call systems for highways, and more have all been powered by photovoltaic cells in regions where installing power lines is challenging.

In addition, they are utilized in satellites, batteries, calculators, watches, televisions, computer card readers, and other portable power sources. They are also used for weather monitoring. In addition to this, PV are utilized to power pump sets for irrigation, drinking water supply, and rural electricity generation that is, street lamps, so forth.

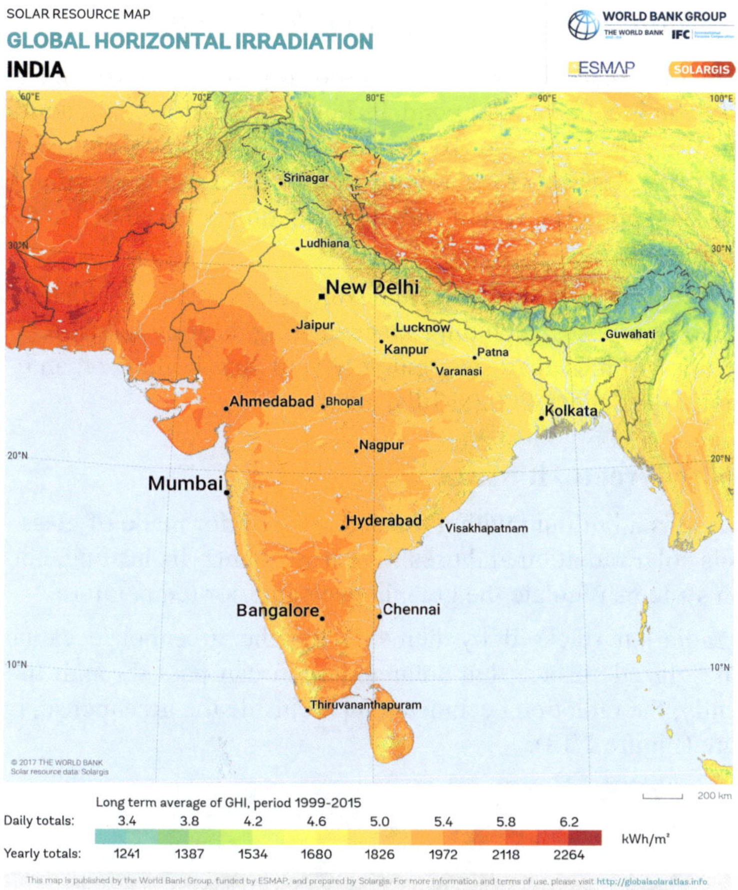

Figure 2.17. *The solar rooftop power generation scheme of the state government is fast gaining popularity among citizens.*

Source: By © The World Bank, Source: Global Solar Atlas 2.0, Solar resource data: Solargis, CC BY 4.0, https://commons.wikimedia.org/w/index.php?curid=74021501.

2.2.8. Solar Thermal Power Production

The process of converting solar energy into electrical power using thermal energy is known as solar thermal power production. In this process, a working fluid—gas, water, or any other volatile liquid—is first heated using solar energy. A turbine is used to transform this thermal energy into mechanical energy. Ultimately, this mechanical energy is transformed into electrical energy by a turbine and a conventional generator.

2.2.8.1. Production of Power Through Solar Ponds

A natural or man-made body of water used to gather, absorb, and store solar radiation as heat is called a solar pond. Its bottom is made of radiation-absorbing black plastic, and it is only 5 to 10 cm deep. It is protected from solar radiation by a curved fiber glass cover that minimizes radiation and convection losses (air movement). The pond is covered with an insulating material bed to reduce heat loss to the ground.

Solar ponds use water to collect and store solar radiation, which can be used for a variety of purposes, including space heating, heating industrial processes, and powering turbines that evaporatively boil low-boiling organic fluids to create electricity (Scheer, 2002; Mills, 2004).

2.2.9. Solar Green Houses

A greenhouse is a building with a transparent exterior made of glass or plastic that collects solar radiation and uses it to grow plants. Its heating, cooling, and ventilation systems regulate the greenhouse's interior temperature.

Heat radiation released by items inside the greenhouse cannot escape through the glazed surface, but solar radiation can pass through the glazing. Consequently, the radiation becomes trapped inside the greenhouse, raising the temperature (Figure 2.18).

Figure 2.18. *A solar greenhouse, on the other hand, creates a warm, nurturing environment all year long using nothing more than natural materials, energy-efficient design and the power of the sun.*

Source: By User:Mattes – Own work, Public Domain, https://commons.wikimedia.org/w/index.php?curid=1128087.

Since there is no air mixing between the greenhouse and surrounding air due to the closed boundary of the greenhouse structure, the air inside the greenhouse becomes enriched with CO_2. Furthermore, because transpiration is restricted, there is less loss of moisture. All of these characteristics support plant growth all year long, not just during the day but also at night.

2.3. SOLAR POWER PLANT TECHNOLOGY

Solar thermal power plants are divided into 5 categories:

1. Parabolic trough.
2. Central receiver system (CRS).
3. Parabolic dish/Sterling dish.
4. Solar chimney.
5. Fresnel collector.

2.3.1. Parabolic Trough Power Plant

This power plant employs long, parallel rows of concentrators. Each concentrator is a glass parabolic reflecting surface that is positioned on an agent structure to create a mirror-like effect. The absorber tubes with special coatings and Pyrex glass placed along the focal line make up the receiver. The primary structural beams support the distal portions of the receiver, which are positioned on the two fulcrums. To ensure that the sun's rays are reflected on the absorber tube for the duration of the tracking time, tracking systems are single-axis devices that track the sun from east to west (Figure 2.19).

Figure 2.19. *Parabolic trough at a plant near Harper Lake.*

Source: https://upload.wikimedia.org/wikipedia/commons/a/a8/Parabolic_trough_at_Harper_Lake_in_California.jpg.

Once it reaches the heat exchanger, a heat transfer fluid (oil) that is approximately 400□ in temperature transforms water into steam. The superheated steam then passes through a turbine to produce electrical energy. Even after dusk, this kind of heat-storage power plant can produce electricity.

The main components of a parabolic power plant include:

- Parabolic reflectors.
- Receiver; to absorb reflected radiation and heat the heat transfer fluid.
- Steel supporter and foundation.
- Electric power production systems.
- Heat transfer equipment.
- Equipment to transfer heat loss to the outside environment.

2.3.2. Central Receiver System (CRS) Power Station

A series of mirrors, or heliostats, are part of this system. They are designed to concentrate solar energy independently and reflect it back to the central receiver located atop the tall tower. The tower-mounted heat exchanger transforms water into superheated steam using the energy gathered in the central receiver. To generate electricity, a generator turbine is installed at the base of the tower and is driven by this steam.

The main components of a CRS power station:

1. Heliostat, solar radiation collector system.
2. Central receiver, which absorbs heat from solar radiation and make it usable.
3. The thermal energy transfer system absorbs heat passed through the receiver and moves it around. In the basic design, water, and steam are used as absorbent and heat transporter fluid, however in the more developed designs, molten sodium and potassium have been used.
4. The power conversion system.
5. Energy storage system.

2.3.3. Sterling Dish Power Plants

Sunlight is directed onto the parabolic concentrator's surface and collected at the focal point to power these plants. The collectors need to follow the sun at all times for these systems to work. Dual-axis tracking mechanisms are therefore required. In these systems, concentrated solar radiation is directed towards the focal point, wherein concentrated thermal energy is transformed into mechanical energy via a Sterling engine. This mechanical energy is then used by an alternator to produce electricity.

The main components of Sterling dish power plants include:

1. Concentrator surfaces that focus the rays of the sun at the focal point.
2. Sterling engine converts concentrated thermal energy into mechanical energy and generates electricity via an alternator.
3. Tracking system and control system track the sun making sure that the receivers are always pointing toward the sun to focus the light on the receiver of the Sterling engine. The control system uses different information coming from different sensors and the Sterling engine to issue the right control signal in any command mode. Structural and foundation: to keep the concentrators, Sterling engines and other components safe from inertia loads, wind, and earthquakes; a strong foundation and structure are needed (Figure 2.20).

Figure 2.20. *10 kW Dish-stirling system.*

Source: https://upload.wikimedia.org/wikipedia/commons/f/f2/Dish-stirling-at-odeillo.jpg.

2.3.4. Solar Chimney Power Plant

This kind of solar power system is the combination of a solar air collector and air steering column that produces an inducted air stream. The air stream causes the turbine to rotate and generate electricity.

2.3.4.1. Application of Solar Chimney Power

The central tower's photoconductive collector roof, which encircles the chimney, is warmed by Plant Sun. The collector roof's junction with the central tower is built to withstand air infiltration. Warm air will ascend through the tower because it is lighter than cold air. Constant air flow drives the turbines to convert air energy into mechanical energy, which is subsequently converted into electrical energy via generators. Under the collector roof, water pipes or containers are used to generate electricity continuously. There is just one filling required for these pipes or containers.

The main components of solar chimney power plant:

- The semi-transparent roof (e.g., glass) is installed at several meters above the ground.
- An elevated chimney is located in the center of the collector roof.
- Wind turbines are installed in the junction of the chimney with the roof.
- Piece of coated land (Figure 2.21).

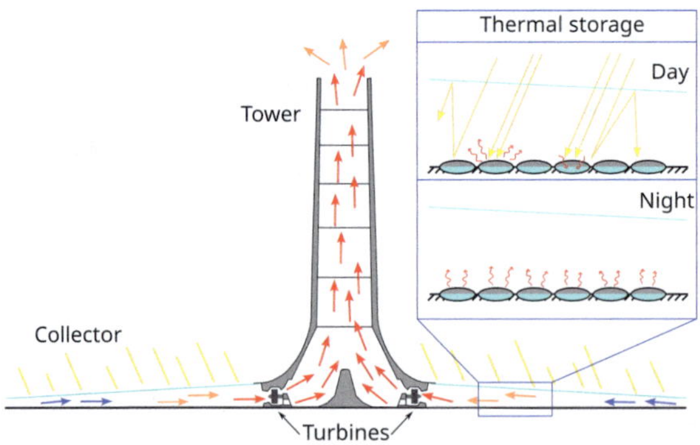

Figure 2.21. *A test solar updraft tower in Manzanares, Spain, ran successfully for several years in the 1980s before toppling in a windstorm. Today, new efforts to build solar chimneys are meeting with financing challenges.*

Source: By Redrawn and slightly modified by Cryonic07. Original jpg-drawing was made by Utilisateur: Kilohn limahn – Own work based on: Schéma fonctionnel.jpg and Original jpg drawing, CC BY-SA 3.0, https://commons.wikimedia.org/w/index.php?curid=2765191.

2.3.5. Fresnel Solar Power Station

The Fresnel collector in this kind of power plant is used to focus sunlight onto the receiver tube. The collectors are installed in a north-south linear order, similar to the parabolic power plant. There are lots of long, narrow, flat mirrors among collectors. Each mirror's angle is adjusted to best focus the receiver's reflection of sunlight. The receiver has a secondary reflector in the form of a parabolic pair to gather reflected energy and focus it on the receiver tube, allowing the fluid (e.g., the receiver tube's oil) will be heated. After passing through a heat exchanger, the heated fluid powers a steam generator (Figure 2.22).

Figure 2.22. *A solar power tower at Crescent Dunes Solar Energy Project concentrating light via 10,000 mirrored heliostats spanning thirteen million sq ft (1.21 km²).*

Source: By Amble – Own work, CC BY-SA 4.0, https://commons.wikimedia.org/w/index.php?curid=37624786.

2.4. IMPACT ON CARBON FOOTPRINT

Because solar energy doesn't emit greenhouse gases while producing electricity, it drastically lowers carbon footprints. In contrast to fossil fuels, which emit carbon dioxide and other harmful emissions, solar power uses sunlight to generate clean energy. While there are some carbon emissions related to the production and installation of solar panels, they pale in comparison to the long-term emissions resulting from the burning of fossil fuels. To reduce overall greenhouse gas emissions and move towards a more sustainable and climate-friendly energy system, solar energy is a very effective tool because, throughout their operational lifespan, solar panels offset their initial carbon footprint multifold. When it comes to lowering the carbon footprint, solar energy outperforms traditional fossil fuels. In contrast to coal, oil, or natural gas, solar power generation uses photovoltaic cells or solar thermal systems to capture sunlight and generate electricity without releasing greenhouse gases into the atmosphere. The fight against climate change and the mitigation of global warming depend heavily on this large cut in carbon emissions. Carbon emissions are produced during the production, shipping, and installation of solar panels, but they are negligible in comparison to the emissions produced for a lifetime from fossil fuel energy sources. The carbon footprint associated with solar energy is significantly reduced over the course of a solar power system's operational life, which is usually 20 to 30 years. Furthermore, these initial emissions are continuously being decreased by technological advancements and improvements in manufacturing processes. Businesses and individuals can drastically reduce their carbon footprint, support cleaner air, and help ensure a more sustainable energy future by switching to solar energy.

2.4.1. Quantifying Emissions Reduction from Solar Energy

Quantifying emissions reduction from solar energy involves comparing the lifecycle carbon emissions of solar power to those of fossil fuels. The lifecycle analysis includes all stages: manufacturing, installation, operation, maintenance, and decommissioning of solar panels. On average, solar photovoltaic (PV) systems produce about 20 to 50 grams of CO_2-equivalent emissions per kilowatt-hour (g CO_2e/kWh) of electricity generated over their lifetime. In contrast, coal-fired power plants emit approximately 800 to 1,000 g CO_2e/kWh, and natural gas plants emit about 450 to 550 g CO_2e/kWh.

For example, depending on the carbon intensity of the local power grid, a standard solar PV system with a capacity of 5 kW might reduce emissions by approximately 4 to 6 metric tons of CO_2 annually. This could lead to a total reduction of 120 to 180 metric tons of CO_2 over a 30-year lifespan. The amount of sunlight received, the carbon intensity of the local electricity grid, and the efficiency of the solar panels are some of the variables that affect the actual reduction. However, when compared to traditional fossil fuel sources, solar energy continuously provides a considerable reduction in greenhouse gas emissions, helping to mitigate climate change and produce cleaner energy.

2.4.2. Practical Considerations: Installation, Cost, and Geographic Suitability

When evaluating solar energy, practical considerations include installation logistics, cost, and geographic suitability:

1. **Installation** involves several steps: assessing site suitability, securing permits, and installing solar panels, inverters, and mounting systems. The site must have adequate sunlight exposure and minimal shading from trees or buildings. Professional installation is crucial for optimizing system performance and ensuring safety and compliance with local regulations.
2. **Cost** is a major factor. The initial investment for solar panels, inverters, and installation can be significant, though prices have decreased over the years due to technological advancements and economies of scale. Financial incentives such as tax credits, rebates, and net metering can offset these costs, making solar energy more accessible. Additionally, while the upfront cost can be high, solar systems typically result in lower electricity bills over time and offer long-term savings.
3. **Geographic Suitability** greatly impacts the efficiency of solar energy systems. Solar panel performance is influenced by geographic location, weather patterns, and seasonal variations in sunlight. Areas with high solar irradiance, such as sunny regions, are more suitable for solar energy. However, advancements in technology have improved the performance of solar panels in less sunny regions as well. For optimal results, it's essential to consider local climate conditions and solar resource availability when planning a solar energy system.

Overall, while solar energy presents an excellent opportunity for reducing carbon footprints and energy costs, careful consideration of installation requirements, financial implications, and geographic factors is essential for maximizing its benefits and achieving a successful and cost-effective solar energy system (Scheer, 2002; Mills, 2004).

REFERENCES

1. Agrafiotis, C., et al. (2005). Solar water splitting for hydrogen production with monolithic reactors. *Solar Energy, 79*(4), 409–421. https://doi.org/10.1016/j.solener.2005.01.004.
2. Denzer, A. (2013). *The Solar House: Pioneering Sustainable Design*. Rizzoli. ISBN 978-0847840052.
3. Kaul, K., et al. (2001). Row orientation affects fruit yield in field-grown okra. *Journal of Sustainable Agriculture, 17*(2/3), 169–174. https://doi.org/10.1300/J064v17n02_16.
4. Leon, M., & Kumar, S. (2007). Mathematical modeling and thermal performance analysis of unglazed transpired solar collectors. *Solar Energy, 81*(1), 62–75. https://doi.org/10.1016/j.solener.2006.02.011.
5. Martin, C. L., & Goswami, D. Y. (2005). *Solar Energy Pocket Reference*. International Solar Energy Society.
6. Meier, A., et al. (2005). Solar chemical reactor technology for industrial production of lime. *Solar Energy, 80*(10), 1355–1362. https://doi.org/10.1016/j.solener.2005.01.018.
7. Mills, D. (2004). Advances in solar thermal electricity technology. *Solar Energy, 76*(1–3), 19–31. https://doi.org/10.1016/j.solener.2003.10.020.
8. Müller, R., & Steinfeld, A. (2007). Band-approximated radiative heat transfer analysis of a solar chemical reactor for the thermal dissociation of zinc oxide. *Solar Energy, 81*(10), 1285–1294. https://doi.org/10.1016/j.solener.2007.04.010.
9. Scheer, H. (2002). *The Solar Economy: Renewable Energy for a Sustainable Global Future*. Earthscan Publications Ltd.
10. Schittich, C. (2003). *Solar Architecture: Strategies, Visions, Concepts*. Architektur-Dokumentation GmbH & Co. KG.
11. Smil, V. (2003). *Energy at the Crossroads: Global Perspectives and Uncertainties*. MIT Press.
12. Smil, V. (2006). *Energy at the Crossroads* [PDF]. Organisation for Economic Co-operation and Development. Retrieved from: https://www.oecd.org (accessed on 9 November 2024).

Chapter **3**

Wind Energy

Contents

INTRODUCTION

Wind energy is a rapidly expanding renewable energy source that produces electricity by using the kinetic energy of the wind. Because it can lessen reliance on fossil fuels and greenhouse gas emissions, this energy source has grown in popularity. The main technology for capturing wind energy is a wind turbine, which rotates blades that are connected to a generator to transform air movement into electrical power. Both onshore and offshore locations are possible for these turbines, each with their own benefits and drawbacks.

Onshore wind farms are frequently located in places with high average wind speeds, like elevated or open plains. Compared to offshore sites, these locations usually offer a simpler installation procedure and less expensive options. Onshore wind farms, however, may encounter difficulties with noise, aesthetic impact, and possible effects on nearby wildlife. These issues are still being addressed by developments in turbine technology and site selection, which increase productivity while reducing environmental disturbance.

Conversely, offshore wind farms take advantage of the stronger and more constant winds that blow over bodies of water. These farms can produce large amounts of energy because they are often located in deeper waters with stronger and more consistent winds. The main obstacles facing offshore wind include greater costs for installation and upkeep as well as technical challenges arising from harsh marine environments. Offshore wind is a very attractive option for increasing renewable energy capacity despite these obstacles because of the possibility of large-scale energy production and the decreased impact on populated areas.

Technological advancements such as better wind pattern prediction models, bigger, and more effective turbines, and energy storage system advancements all contribute to the growth of wind energy. These developments strengthen wind power's dependability and economic feasibility, positioning it as a crucial element of international climate change mitigation plans. Additionally, by generating jobs in manufacturing, installation, and maintenance, wind energy projects frequently boost local economies.

But switching to wind energy is not without its difficulties. Wind speed variation can result in uneven power production, which calls for the creation of complementary energy storage technologies or grid improvements. Furthermore, careful planning and infrastructure upgrades are needed to integrate significant amounts of wind power into the current energy grids. A viable and increasingly effective option for producing electricity sustainably is wind energy. With the development of technology and improved integration strategies, wind energy is expected to become increasingly important in the global drive toward cleaner, more dependable energy systems (European Wind Energy Association, 2009; Nelson, 2009).

3.1. CONCEPT OF WIND ENERGY

Air is moved by wind, which is brought on by variations in atmospheric pressure. The earth's surface soil warms as the sun shines on it. The air above the soil is thus warmed by it. Warm air rises like a hot-air balloon because it is less dense than cool air. As cool air replaces it, heated air flows in. The convection cycle is eventually completed when the warm air that is rising cools and descends back to Earth. As long as the solar engine powering it is in the sky, this cycle is repeated repeatedly, rotating like a car's crankshaft. Heat is carried around the world by the atmosphere, which functions as a massive solar-powered engine. Heat is transferred from lower latitudes to northern regions by the large-scale convection currents that are sparked by the sun's rays.

The flows of air that rush across the surface of the earth in response to this global circulation are called wind. This wind resource is renewable and inexhaustible, as long as sunlight reaches the earth (Figure 3.1).

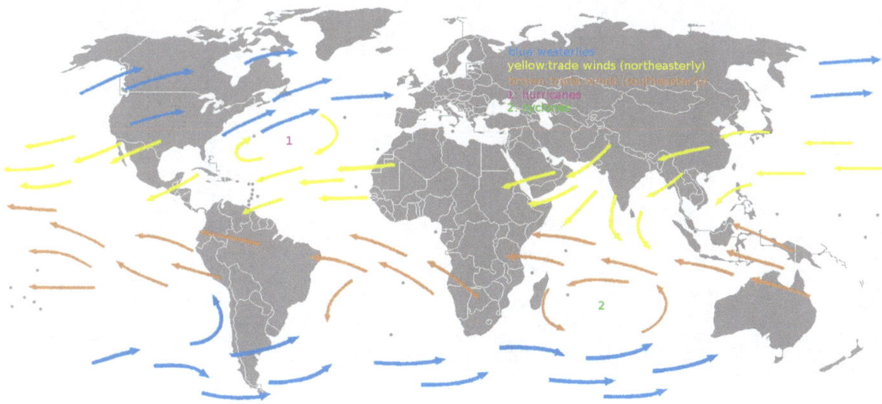

Figure 3.1. *Wind movement because of pressure deference.*

Source: By KVDP – Own work, Public Domain, https://commons.wikimedia.org/w/index.php?curid=8158719.

3.1.1. Importance of Wind Energy

Wind energy, a form of solar energy that is produced by the movement of air relative to the Earth's surface. This form of energy is generated by the uneven heating of Earth's surface by the Sun and is modified by Earth's rotation and surface topography.

One major participant in the global energy market and a common energy source for new power generation is wind energy. The technical maturity and rapid deployment of wind power, as well as the fact that the percentage of wind that can be integrated into the electrical system has no practical upper limit, are acknowledged as attributes of this leading energy technology. An approximate value of 1.08×10^{11} MW is the total solar power that the earth receives. Just 2% (i.e.,) of this solar input was received (i.e., 3.6×10^9 MW) is transformed into wind energy, with roughly 35% of that energy dissipating within a 1000 m radius around the surface of the planet. Thus, there is about 1.26×10^9 MW of available wind power that can be transformed into other types of energy. Wind energy has the potential to supply all of the world's energy needs because this figure is 20 times faster than the rate at which energy is currently consumed worldwide (Figure 3.2).

Figure 3.2. *Wind power in the UK.*

Source: By 林 慕*/Chris Lim from East Coast (*海岸*), Singapore (*新加坡*) – Windmills in China?{D70 series}, CC BY-SA 2.0, https://commons.wikimedia.org/w/index.php?curid=2909338.*

Wind energy has several advantages and benefits over conventional energy sources. Wind power is a clean and environmentally friendly energy source, in contrast to nuclear power, which produces radioactive waste, and fossil fuels, which release harmful gases into the atmosphere. It is abundant and available in most parts of the world as a free, limitless energy source. Furthermore, increasing the use of wind power would lessen the need for fossil fuels, which, based on current consumption rates, may run out sometime this century. Moreover, wind energy is far less expensive per kWh than solar energy. Therefore, it is anticipated that wind energy, the most promising energy source, will be essential to the world's power supply in the twenty-first century.

3.1.1.1. Coriolis Force

During the day, the air above the land heats up more quickly than the air over water. The warm air over the land expands and rises, and the heavier, cooler air rushes in to take its place, creating winds. At night, the winds are reversed because the air cools more rapidly over land than over water. This wind flow is called local wind.

3.1.1.2. Power Available in Wind

The power that can be captured from the wind with a wind turbine with effective Area Ar is given by:

Power available in wind (in Watts) = ½ * air density * swept area * wind velocity

P_a =1/2 *P*A*V or P_a =1/8 *P* D *V

where, air density P = 1.23 kg per cubic meter at sea level, and the swept area is in square meters and wind velocity is in meters per second.

The only way to increase the available power in low winds is by sweeping a larger area with the blades and that's the 2nd key concept from this formula. When the diameter formed by rotation of the blades doubles, power available in wind increases by a factor of four. However, there's no way to harvest ALL of this available energy and turn it into electricity.

In 1919 a gentleman named Betz calculated that there's a limit to how much power a turbine blade can extract from the wind. Beyond the Betz Limit of 59.26% energy extraction, more, and more air tends to go around the turbine rather than through it, with air pooling up in front. So 59.26% is the absolute maximum.

There are more losses following Betz. Even when operating at their optimal speed, small wind turbine blades are never 100% efficient. It is impossible for an alternator or generator to produce electricity from a rotating shaft with 100% efficiency. Friction losses occur from any gearing used in the power conversion process as well as from bearings. Additionally, the alternator or generator experiences losses from electrical resistance and magnetic drag (Figure 3.3).

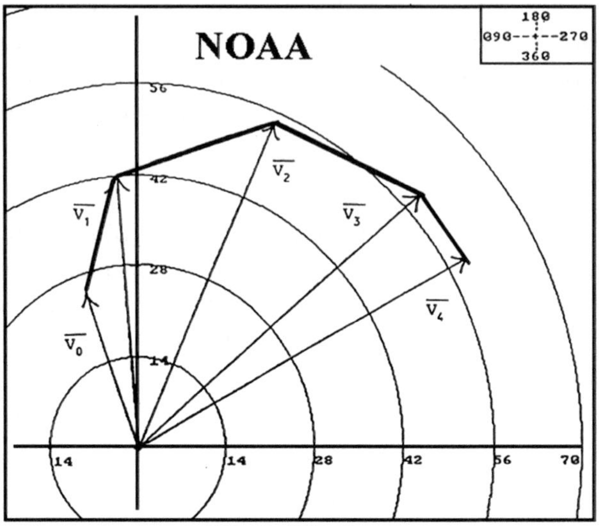

Winds / Vents

Height / Hauteur (km)	Dir	Speed / Vitesse
Sfc	161°	25
1	175°	42
2	204°	54
3	229°	59
4	241°	59

Figure 3.3. *A hodograph.*

Source: By Local Analysis and Prediction System of NOAA – http://laps.noaa.gov/afwa/eds_stuff/introfig_hodo.jpg from this article with modifications to make it French and English with MSPaint, Public Domain, https://commons.wikimedia.org/w/index.php?curid=1375015.

Considering the Betz limit, the maximum power output from a 40-meter diameter Turbine in a 10 m/s wind speed is calculated as follows: Power available (Watts)

$P_{max} = 0.5926 * ½ * 1.23 * 1256 * 10 = 457{,}747$ Watts = 0.457 MW

A perfect turbine|| would work right at the Betz limit, the blades and the alternator would match perfectly at all wind speeds, and the alternator would have no internal magnetic or electrical losses. This is also known as the coefficient of power Cp. Power co-efficient (Cp), describes the fraction of the power in the wind that may converted by the wind turbine into mechanical work.

$$Cp = \frac{Power\ Output\ from\ Wind\ Machine}{Power\ avilable\ in\ wind}$$

It is the fraction of power in a wind stream that can be extracted. It has a theoretical maximum value of:

Cp (max) = 0.593 (popularly known as Betz co-efficient)

The actual power output of a wind turbine is given by

3 P = 0.5 x P x A x Cp x V x x g b

where:

- P = Power in watts (746 watts = 1 hp) (1,000 watts = 1 kilowatt)
- P = Air density (about 1.225 kg/m^3 at sea level, less high up)
- 2 A = Rotor swept area, exposed to the wind (m)
- Cp = Power Coefficient [0.59 {Betz limit}is the maximum theoretically possible, 0.35 for a good design]
- V = Wind speed in meters/sec (20 mph = 9 m/s)
- η_g= Generator efficiency (80% or possibly more for a permanent magnet generator or grid-connected induction generator)
- η_g= Gearbox/bearings efficiency (could be as high as 95% if good)

The actual power conversion efficiency (Coefficient of Power or Cp) of a wind turbine is of the order of 0.20–0.35 (Figure 3.4). Considering a Cp value of 0.3 for a turbine, the power output from a 40-meter diameter turbine in a 10 m/s wind speed will be P = 0.2319 MW act

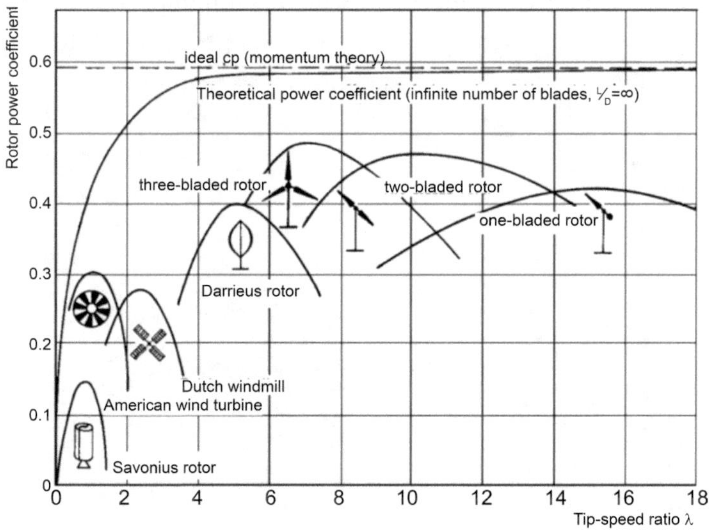

Figure 3.4. *A power coefficient that depends on λ and the type of wind turbine.*

Source: Castillo OC, Andrade VR, Rivas JJR, Gonzălez RO. Comparison of Power Coefficients in Wind Turbines Considering the Tip Speed Ratio and Blade Pitch Angle. Energies. 2023; 16(6):2774. https://doi.org/10.3390/en16062774.

3.1.2. Benefits of Wind Energy

Numerous factors, such as the larger context of energy supply and demand, the increased awareness of environmental issues, particularly climate change, and the remarkable advancements in the technology itself, are propelling the growth of the wind energy market. In many parts of the world, these elements have come together to promote political support for the growth of the industry.

3.1.2.1. Security of Supply

The rate at which the world's energy needs are growing is astounding, necessitating large investments in new power plants and grid infrastructure. The demand for energy is rising at the same time that fossil fuel supplies are running low and prices are at an all-time high. On the other hand, wind energy is a plentiful, native power source that can be found almost anywhere in the world.

The world's energy needs are expected to increase by 30 to 60% by 2030, depending on the efficiency measures put in place. According to estimates from the International Energy Agency (IEA), more than US$ 13 trillion in investments will be needed to install 4500 GW of new energy capacity before 2030. The world's energy demand is expected to rise sharply, necessitating large investments in new power-generating capacity and grid infrastructure, particularly in emerging economies like China and India (Figure 3.5).

Figure 3.5. *Major tech companies are under pressure to turn to green energy sources to power data centers.*

Source: By Erik Wilde from Berkeley, CA, USA – harvesting wind, CC BY-SA 2.0, https://commons.wikimedia.org/w/index.php?curid=51105579.

Industrialized countries face a different but parallel situation. While demand is increasing, the days of overcapacity in electricity production are coming to an end. Many older power plants will soon reach the end of their working lives. The IEA predicts that by 2030, over 1600 GW of power generation capacity will need to be built in the Organization for Economic Co-operation and Development (OECD) countries alone, including the replacement of retiring plants.

The primary fossil fuels used in power generation are getting harder and more expensive to extract, coinciding with the ongoing rise in energy demand. As a result, a number of the world's largest economies are depending more and more on fuel imports, which can come from expensive places where the security of the fuel supply is threatened by political unrest and conflict. Wind energy is a massive indigenous power source that is permanently available in almost every country in the world, in contrast to the unpredictable prices and uncertain supply of conventional fuels.

3.1.2.2. Economic Considerations

Economically speaking, wind energy makes sense. Unlike other forms of energy, the cost of fuel required for a wind turbine to run its entire life is known to be zero. This eliminates a sizable portion of the risk for the investor. Even with newly developed conventional technologies, wind power is already competitive and sometimes even more affordable. When one considers the cost of carbon, wind power becomes even more appealing. Another important positive factor in the economic analysis of wind power is the creation of jobs and regional economic development.

3.1.2.3. No Fuel Price Risk

Wind energy is a massive indigenous power source that is permanently available in almost every country in the world, in contrast to the unpredictable prices and uncertain supply of conventional fuels. The volatility of fuel price developments is a major risk factor for conventional generation technologies; in the last year alone, oil prices have fluctuated between 50 and 150 USD. On the other hand, owners of wind farms are aware of the cost of the electricity they produce. Except hydropower, which is recognized as a "renewable power generating technology," no conventional technology can make such a claim. This is a serious concern for government planners trying to lessen their vulnerability to macroeconomic shocks brought on by the whims of global commodities markets, as well as for individual utilities and power plant operators (European Wind Energy Association, 2009; Nelson, 2009).

3.1.3. The Cost of Wind Energy

When compared to newly developed conventional technologies, wind power is already competitive and sometimes even more affordable. Wind power is commercially attractive, especially when considering the price of carbon, which is a factor in an increasing number of markets, even though nothing can match the current, embedded conventional generation plant that has already been paid off (and was mostly constructed with significant state subsidies: governments still subsidize conventional technologies at the rate of about 250 billion USD/year) (Figure 3.6).

Reducing our economies' exposure to fuel price volatility is one of the wind power's most significant economic advantages. This benefit is so great that, even if wind energy were more expensive per kWh than other forms of power generation, it could easily justify a larger share of wind energy in most European countries. Public authorities have been using standard methods for calculating energy costs for over a century, but these methods currently do not account for the risk reduction from wind energy. On the contrary, the way that calculations are currently done clearly favors using high-risk options for producing electricity. This issue needs to be addressed right away because the industrialized world is becoming more and more dependent on importing fuel at erratic and higher prices from politically unstable regions.

Figure 3.6. *If we keep subsidizing wind, will the cost of wind energy go down?*

Source: By Tedder – Own work, CC BY 3.0, https://commons.wikimedia.org/w/index.php?curid=6697607.

Because markets do not fairly value the external effects of power generation, markets cannot resolve these issues on their own. Because the costs and benefits for a household or business that purchases or sells in the market differ from the costs and benefits to society, governments must address market failures brought about by external effects. Power companies can dispose of their waste more affordably, for example, in the form of free methane, fly ashes, CO_2, nitrous oxides, and sulfur oxides. The issue is that it costs other people money, for example, in the shape of lung conditions, acid rain damage, or global warming. Similar to this, employing wind energy benefits society and the economy overall rather than specific market players.

3.1.3.1. Speed of Deployment

Another benefit of wind power is that it can be implemented more quickly than other energy supply technologies. A conventional power plant can take 10 or 12 years or longer to build, and until it is finished, it cannot produce electricity. A partially constructed wind farm is essentially a smaller power plant that begins to produce electricity and revenue as soon as the first turbine is connected to the grid. The deployment of wind power is measured in months. Given the imminent threat of climate change, the ability to install large offshore wind farms—which demand a higher degree of infrastructure and grid network connection—in less than two years is an essential advantage. For conventional thermal, nuclear, or hydropower plants, the project execution cycle typically lasts four to five years.

3.1.3.2. Emissions Free

Climate change is now generally accepted to be the greatest environmental threat facing the world, and keeping our planet's temperature at sustainable levels has become one of the major concerns of policy makers. The United Nation's (UN) Intergovernmental Panel on Climate Change (IPCC) projects that average temperatures around the world will increase by up to 5.8□ over the coming century. This is predicted to result in a wide range of climate shifts, including melting ice caps, flooding of low-lying land, storms, droughts, and violent changes in weather patterns (Figure 3.7).

The electricity industry is the biggest source of emissions, contributing roughly 25% of total emissions and 40% of CO_2 emissions, even though it is by no means the only cause of climate change. Between now and 2020, there are essentially three ways to significantly reduce emissions in the power sector: energy conservation and efficiency; switching from coal to gas as a fuel; and renewable energy, mostly wind power.

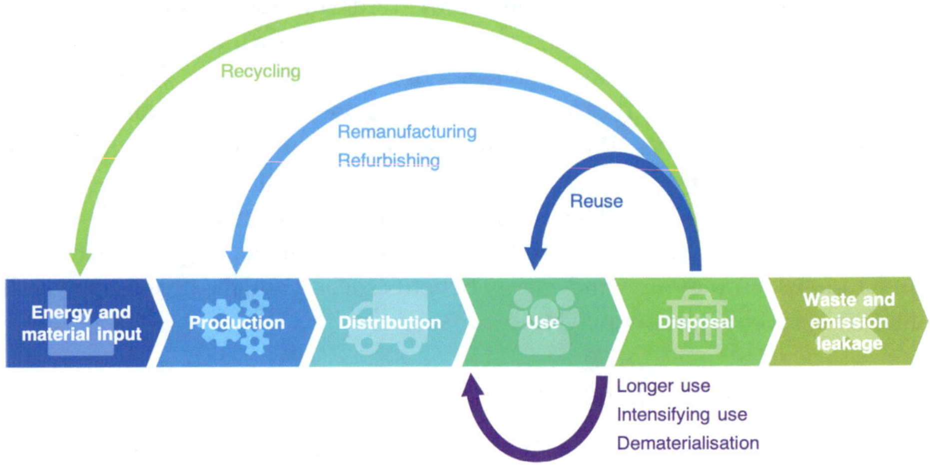

Figure 3.7. *The energy system works in a circular manner.*

Source: By Geissdoerfer, M., Pieroni, M.P., Pigosso, D.C. and Soufani, K. – Geissdoerfer, M., Pieroni, M.P., Pigosso, D.C. and Soufani, K., 2020. Circular business models: A review. Journal of Cleaner Production, p.123741, CC BY 4.0, https://commons.wikimedia.org/w/index.php?curid=97396208.

Because wind energy doesn't release any of the other air pollutants or carbon dioxide that contribute to climate change, it doesn't have the high external costs associated with using conventional energy sources. In just three to six months of operation, a wind turbine can run carbon-free for the remaining 20 years of its 20-year lifespan, having offset all emissions caused during construction. Wind power is risk-free insurance against the long-term downside of carbon-intensive investments in a world where carbon emissions are becoming more and more restricted.

3.1.3.3. Remedy for Environmental Concerns

One clean, emission-free method of generating electricity is wind power. It is dependent on harnessing the energy of the natural world, just like all renewable energy sources, and it doesn't release any pollutants like conventional fuels do. In addition to offering many other environmental advantages, wind energy is the only power generation technology capable of achieving the significant reductions in CO_2 emissions required to fend off the worst effects of climate change. Because it emits fewer harmful air pollutants than other generation technologies, it helps reduce air pollution, which is choking cities all over the world. Toxic waste is not produced by wind energy. Furthermore, wind energy uses almost no water, which is a significant environmental factor in a world where water is becoming more scarce.

3.1.3.4. Combating Climate Change

First and foremost, when wind energy is in operation, it emits zero carbon dioxide, the primary greenhouse gas linked to climate change, and very little when its equipment is made and wind farms are built. In contrast, carbon dioxide emissions from fossil fuels like coal, gas, and oil are significant. Wind energy is expected to save 10 billion tons of CO_2 by 2020, making it a vital weapon in the fight against climate change.

3.1.3.5. Water Conservation

Another consideration of wind energy deployment concerns water. In an increasingly water-stressed world, wind power uses virtually none of this most precious of commodities in its operation. Most conventional technologies, from mining and extraction to fuel processing and plant cooling, measure their water use in the millions of liters per day (Figure 3.8).

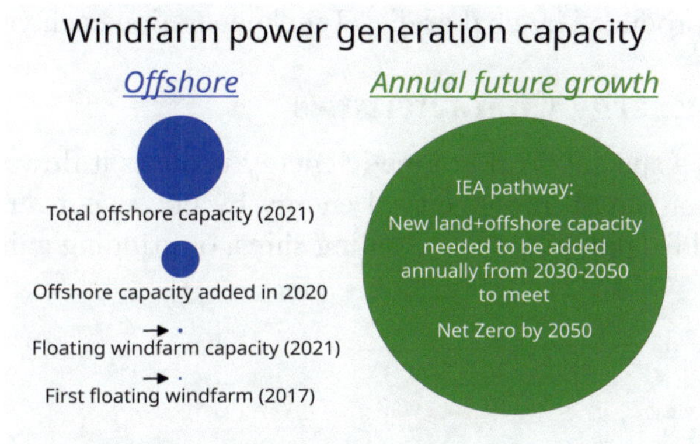

Figure 3.8. *The emerging offshore wind energy industry.*

Source: By RCraig09 – Own work, CC BY-SA 4.0, https://commons.wikimedia.org/w/index.php?curid=109499308.

Other environmental effects resulting from the range of fuels currently used to generate electricity include the landscape degradation and dangers of fossil fuel exploration and mining, the pollution caused by accidental oil spills and the health risks associated with radiation produced by the routine operation and waste management of the nuclear fuel cycle. Exploiting renewable sources of energy, including wind power, avoids these risks and hazards.

Wind energy is an ideal renewable energy because:

1. It is a pollution-free, infinitely sustainable form of energy.
2. It doesn't require fuel.

3. It doesn't create greenhouse gasses.
4. It doesn't produce toxic or radioactive waste.
5. Wind energy is quiet and does not present any significant hazard to birds or other wildlife.
6. When large arrays of wind turbines are installed on farmland, only about 2% of the land area is required for the wind turbines. The rest is available for farming, livestock, and other uses.
7. Landowners often receive payment for the use of their land, which enhances their income and increases the value of the land.
8. Ownership of wind turbine generators by individuals and the community allows people to participate directly in the preservation of our environment.
9. Each megawatt-hour of electricity that is generated by wind energy helps to reduce the 0.8 to 0.9 tons of greenhouse gas emissions that are produced by coal or diesel fuel generation each year.

3.1.4. Wind Energy Characteristics

Wind energy is a special form of kinetic energy in air as it flows. Wind energy can be either converted into electrical energy by power-converting machines or directly used for pumping water, sailing ships, or grinding gain (Figure 3.9).

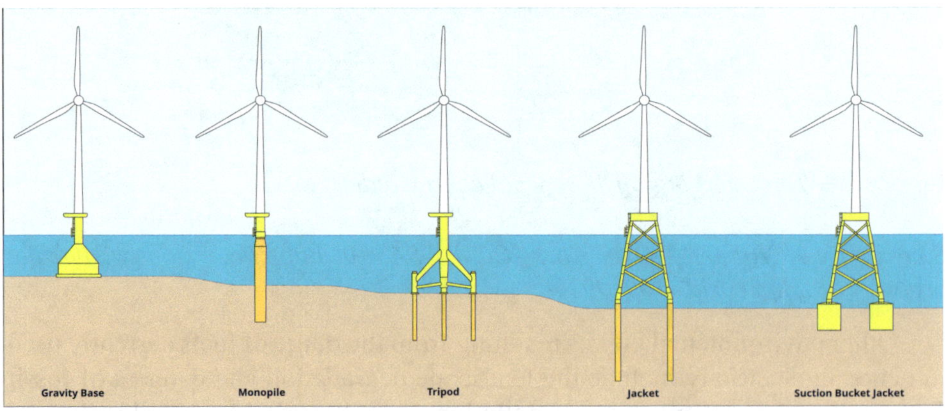

Figure 3.9. *Offshore systems.*

Source: By Ghost456 – Own work, CC BY-SA 4.0, https://commons.wikimedia.org/w/index.php?curid=116764131.

3.1.4.1. Wind Power

Kinetic energy exists whenever an object of a given mass is in motion with a translational or rotational speed. When air is in motion, the kinetic energy in moving air can be determined as:

$$E_k = \tfrac{1}{2} m\overline{u}^2 \quad (1)$$

where, m is the air mass and $\overline{u}$ is the mean wind speed over a suitable time period. The wind power can be obtained by differentiating the kinetic energy in wind with respect to time, that is:

$$P_w = \frac{dE_k}{dt} = \frac{1}{2} \dot{m}\overline{u}^2 \quad (2)$$

However, only a small portion of wind power can be converted into electrical power. When wind passes through a wind turbine and drives blades to rotate, the corresponding wind mass flow rate is:

$$\dot{m} = \rho A \overline{u} \quad (3)$$

where ρ is the air density and A is the swept area of blades. Substituting Eq. (3) into Eq. (2), the available power in wind P_w can be expressed as:

$$P_w = \tfrac{1}{2} \rho A \overline{u}^3 \quad (4)$$

An examination of Eq. (4) reveals that to obtain a higher wind power, it requires a higher wind speed, a longer length of blades for gaining a larger swept area, and a higher air density. Because the wind power output is proportional to the cubic power of the mean wind speed, a small variation in wind speed can result in a large change in wind power (Figure 3.10).

Figure 3.10. *Blue H technologies – world's first floating wind turbine.*

Source: By Green Storm 7 – Own work, CC BY-SA 4.0, https://commons.wikimedia.org/w/index.php?curid=57045610.

3.1.4.2. Blade Swept Area

The blade swept area can be calculated from the formula:

$$A = \pi\left[(l+r)^2 - r^2\right] = \pi l\,(l+2r) \tag{5}$$

where l is the length of wind blades and r is the radius of the hub. Thus, by doubling the length of wind blades, the swept area can be increased by a factor up to 4. When $l \gg 2r$, $A \approx \pi l^2$.

3.1.4.3. Air Density

Another important parameter that directly affects the wind power generation is the density of air, which can be calculated from the equation of state:

$$\rho = \frac{p}{RT} \tag{6}$$

where p is the local air pressure, R is the gas constant (287 J/kg-K for air), and T is the local air temperature in K.

The hydrostatic equation states that whenever there is no vertical motion, the difference in pressure between two heights is caused by the mass of the air layer:

$$\mathrm{d}p = -\rho g\,\mathrm{d}z \tag{7}$$

where g is the acceleration of gravity. Combining Eq. (6) and Eq. (7), yields

$$\frac{\mathrm{d}p}{p} = -\frac{g}{RT}\mathrm{d}z \tag{8}$$

The acceleration of gravity g decreases with the height above the earth's surface z:

$$g = g_0\left(1 - \frac{4z}{D}\right) \tag{9}$$

where g_0 is the acceleration of gravity at the ground and D is the diameter of the earth. However, for the acceleration of gravity g, the variation in height can be ignored because D is much larger than 4 z.

In addition, temperature is inversely proportional to the height. Assume that d T/d z = c, it can be derived that

$$p = p_0\left(\frac{T}{T_0}\right)^{-g/cR} \tag{10}$$

where p_0 and T_0 are the air pressure and temperature at the ground, respectively. Combining Eq. (6) and Eq (10), it gives

$$\rho = \rho_0 \left(\frac{T}{T_0} \right)^{-(g/cR+1)} = \rho_0 \left(1 + \frac{cz}{T_0} \right)^{-(g/cR+1)} \qquad (11)$$

This equation indicates that the density of air decreases nonlinearly with the height above the sea level.

3.1.4.4. Wind Power Density

Wind power density is a comprehensive index in evaluating the wind resource at a particular site. It is the available wind power in airflow through a perpendicular cross-sectional unit area in a unit time period. The classes of turbines are listed in Table 3.1.

Some of wind resource assessments utilize 50 m towers with sensors installed at intermediate levels (10 m, 20 m, etc.). For large-scale wind plants, a class rating of 4 or higher is preferred.

Table 3.1. *Turbine Specifications*

Floating Wind Turbine and Corresponding Parameters

Project	**Foundation/Turbine Type**	**Time**	**Unit Capacity**
Hywind Tampen	spar type Siemens (Zamudio, Spain)	August 2023	8.6 MW
Wind Float Atlantic	semi-submersible type MHI Vestas V164-8.4 MW (Aarhus, Denmark)	2020	8.4 MW
FLOATGEN	Barge type V164-10.0 MW (Aarhus, Denmark)	2018	10.0 MW
SBM Offshore and IFP Energies Nouvelles	Tension leg type SWT-8.0-154 (Zamudio, Spain)	2020	8.0 MW

Large Offshore Wind Turbines and Parameters

Production Company	**Turbine Type**	**Time**	**Unit Capacity**	**Blade Length**
Siemens Gamesa (Zamudio, Spain)	SG14-222DD	May 2020	14 MW	108 m
General Electric (Boston, MA, USA)	Haliade-X	2020	14 MW	107 m
Vestas (Aarhus, Denmark)	V236-15.0 MW	February 2021	15 MW	115.5 m
Goldwind (Beijing, China)	GWH252-16 MW	June 2023	16 MW	123 m
Mingyang (Zhongshan, China)	MySE 16.0-242	August 2021	16 MW	118 m

Fixed Wind Turbines and Corresponding Parameters

Project	**Type**	**Time**	**Unit Capacity**
Huaneng Cangnan No. 2	monopile type	2023	8.5 MW
Ormonde	tripod type	2012	5.0 MW
Rudong	jacket type	2010	2.5 MW

Floating Wind Turbines in China and Corresponding Parameters

Project	**Foundation Type**	**Time**	**Unit Capacity**	**Impeller Diameter**
Yinling	semi-submersible type	2021	5.5 MW	158 m
Fuyao	semi-submersible type	2022	6.2 MW	152 m
Guanlan	semi-submersible type	2023	7.25 MW	158 m
GuonengGongxiang	semi-submersible type	2023	4.0 MW	130 m

Source: Su X, Wang X, Xu W, Yuan L, Xiong C, Chen J. Offshore Wind Power: Progress of the Edge Tool, Which Can Promote Sustainable Energy Development. Sustainability. 2024; 16(17):7810. https://doi.org/10.3390/su16177810.

3.1.5. Wind Characteristics

Wind varies with the geographical locations, time of day, season, and height above the earth's surface, weather, and local landforms. An understanding of the wind characteristics will help optimize wind turbine design, develop wind measuring techniques, and select wind farm sites.

3.1.5.1. Wind Speed

Wind speed is typically judged as the velocity of the wind. Most measurements of air movement are taken of outside air, and there are several factors that can affect it. Average wind speed is often determined by an anemometer and is usually categorized in a standardized measurement scale, called the Beaufort Scale (Figure 3.11).

Of the major factors that influence wind speed, the most important is called the pressure gradient, created by a graduated disparity in atmospheric pressure that occurs in different places. Some areas have low pressure, while others have higher pressure. For example, a valley may have a higher atmospheric pressure than the peak of a mountain that is only a few miles away. Usually, the pressure increases gradually between both points.

For the most part, air moves along these pressure gradients from high pressure to low pressure. The movement is the major force that creates wind on Earth. The greater the difference in pressure, the greater the wind speed. Therefore, areas that experience a large change in pressure over a short distance will typically have higher wind velocities than those where the change is more gradual.

Another factor that may affect the speed of the wind is local weather conditions. Storm fronts often contribute to air currents, as they can create pressure gradients for the wind to travel along. Also freak storms, like hurricanes or cyclones, can drastically alter wind speed.

Another influence on wind velocity is the presence of Rossby waves. These upper atmospheric air currents manipulate weather patterns in the air below. They are caused by the Coriolis effect. A Rossby wave may influence pressure gradients and stir up faster speeds.

Figure 3.11. *An anemometer is commonly used to measure wind speed.*

Source: https://upload.wikimedia.org/wikipedia/commons/8/81/Weather_Station_at_Feilding_Waste_Water_Treatment_Plant.JPG.

An anemometer is the instrument most frequently used to measure wind speed. The first anemometers had a horizontal spoke wheel mounted on a vertical shaft. The little cups that each spoke has at its end catch the wind and cause the wheel to spin.

The number of times the wheel rotates in a specific period can be used to calculate the wind speed. This is still how a lot of these devices are manufactured. There have also been developments in other forms of anemometers.

Lasers are used by Laser Doppler anemometers to measure wind speeds. Anemometers for windmills work by pointing a fan toward the wind. Hot wire anemometers measure wind speed by measuring the friction created by the wind on an electrically charged wire.

The Beaufort Scale is a standardized measurement of wind speed. It is an empirical rating system originally based on the appearance and height of waves at sea. The system has developed to also include speed ratings for each level in knots, miles per hour, and kilometers per hour.

3.1.5.2. Weibull Distribution

The variation in wind speed at a particular site can be best described using the Weibull distribution function, which illustrates the probability of different mean wind speeds occurring at the site during a period. The probability density function of a Weibull random variable $\overline{u}$ is:

$$f(\overline{u},k,\lambda)=\begin{cases}\frac{k}{\lambda}\left(\frac{\overline{u}}{\lambda}\right)^{k-1}\exp\left(-\left(\frac{\overline{u}}{\lambda}\right)^{k}\right) & \overline{u}\geq 0\\ 0 & \overline{u}<0\end{cases} \quad (12)$$

where l is the scale factor which is closely related to the mean wind speed and k is the shape factor which is a measurement of the width of the distribution. These two parameters can be determined from the statistical analysis of measured wind speed data at the site. It has been reported that Weibull distribution can give good fits to observed wind speed data.

3.1.5.3. Wind Turbulence

Wind turbulence is the fluctuation in wind speed in short time scales, especially for the horizontal velocity component. The wind speed u (t) at any instant time t can be considered as having two components: the mean wind speed $\overline{u}$ and the instantaneous speed fluctuation u′(t), that is (Figure 3.12):

$$u(t)=\overline{u}+u'(t) \quad (13)$$

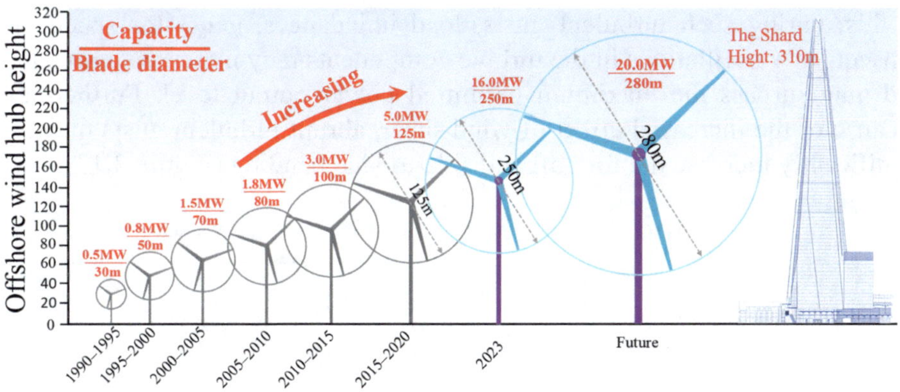

Figure 3.12. *Looking at wind energy production.*

Source: Su X, Wang X, Xu W, Yuan L, Xiong C, Chen J. Offshore Wind Power: Progress of the Edge Tool, Which Can Promote Sustainable Energy Development. Sustainability. 2024; 16(17):7810. https://doi.org/10.3390/su16177810.

Wind turbulence has a strong impact on the power output fluctuation of wind turbines. Heavy turbulence may generate large dynamic fatigue loads acting on the turbine and thus reduce the expected turbine lifetime or result in turbine failure.

In the selection of wind farm sites, the knowledge of wind turbulence intensity is crucial for the stability of wind power production. The wind turbulence intensity I is defined as the ratio of the standard deviation σ_u to the mean wind velocity $\overline{u}$:

$$I = \frac{\sigma_u}{\overline{u}} \tag{14}$$

where both σ_u and $\overline{u}$ are measured at the same point and averaged over the same period.

3.1.5.4. Wind Gust

A wind gust is a phenomenon in which there is a sudden and significant increase in wind speed within a brief period. Wind shear, turbulence, and speed can all change significantly during abrupt, turbulent gusts. During such abrupt turbulent gusts, reducing rotor imbalance and keeping the wind turbine generator's power output constant require relatively quick adjustments to the blades' pitch angle. However, due to the dynamics of the pitch control actuator and the significant inertia of the mechanical components, there is usually a lag between the occurrence of a turbulent gust and the actual pitching of the blades. Because of this, during such turbulent gusts, load imbalances, generator speed, and consequently oscillations in the turbine components may increase significantly and may surpass the maximum permitted power output level. Furthermore, because of the increased effect of wind shear, abrupt turbulent gusts may also significantly increase the fore-aft and side-to-side bending (Figure 3.13).

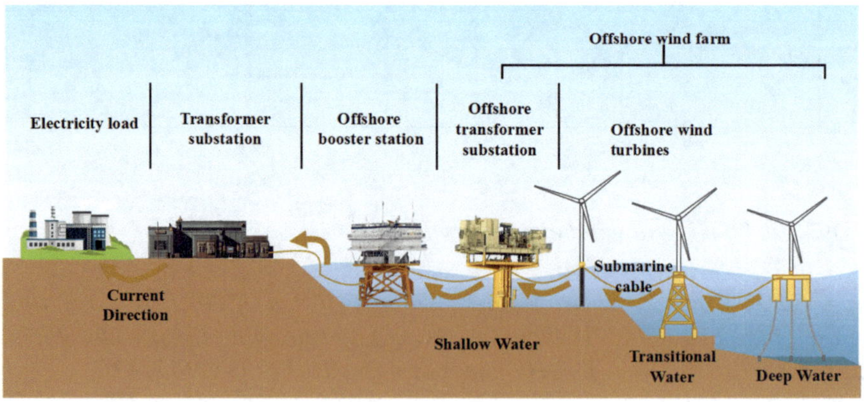

Figure 3.13. *Structural details.*

Source: Su X, Wang X, Xu W, Yuan L, Xiong C, Chen J. Offshore Wind Power: Progress of the Edge Tool, Which Can Promote Sustainable Energy Development. Sustainability. 2024; 16(17):7810. https://doi.org/10.3390/su16177810.

Wind gust predictions are highly desirable to ensure the safe operation of wind farms. Numerous techniques for predicting gusts have been put forth. In contrast to the majority of operational weather forecasting techniques, Brasseur created a novel method for predicting wind gusts that takes physical factors into account. According to a different study, wind gust speeds could be accurately predicted by utilizing a gust factor, which is calculated as the peak gust over the mean wind speed. These findings concur with other researchers' earlier research.

3.1.5.5. Wind Direction

One of the characteristics of the wind is its direction. When choosing a wind farm site and designing the turbine layout, statistical data on wind directions over an extended duration is crucial. When examining wind data related to wind directions at a specific location over a given time period (year, season, month, week, etc.), the wind rose diagram is a helpful tool). The relative frequency of wind directions in eight or 16 major directions is shown in this circular diagram. The wind rose diagram has 16 radial lines that are 22.5° apart from one another. Each line's length varies in direct proportion to the frequency and direction of the wind. The central circle displays a number representing the frequency of calm or nearly calm air (Figure 3.14).

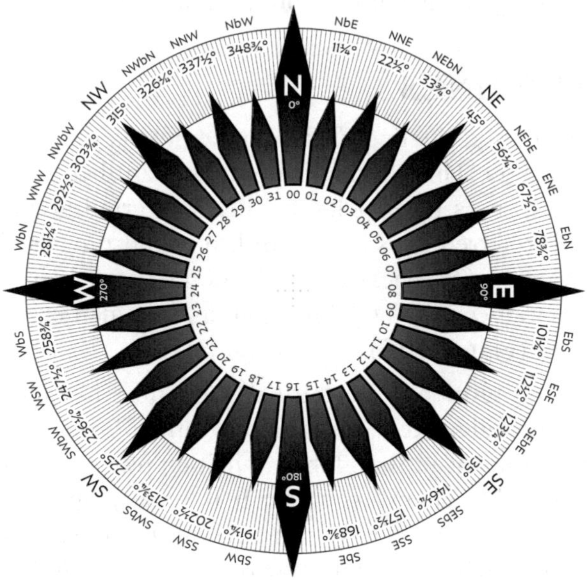

Figure 3.14. *Compass direction directions.*

Source: By JamesLucas – Own work, CC BY-SA 4.0, https://commons.wikimedia.org/w/index.php?curid=117908742.

3.1.5.6. Wind Shear

Wind shear is a meteorological phenomenon in which wind increases with the height above the ground. The effect of height on the wind speed is mainly due to roughness on the earth's surface and can be estimated using the Hellmann power equation that relates wind speeds at two different heights:

$$u(z) = u(z_0)\left(\frac{z}{z_0}\right)^a \tag{15}$$

where z is the height above the earth's surface, z_0 is the reference height for which wind speed u (z_0) is known, and a is the wind shear coefficient. In practice, a depends on a number of factors, including the roughness of the surrounding landscape, height, time of day, season, and location. The wind shear coefficient is generally lower in daytime and higher at night. Empirical results indicate that wind shear often follows the "1/7 power law" (i.e., a = 1/7). The values of wind shear coefficient for different surface roughness are provided. Because the power output of a wind turbine strongly depends on the wind speed at the hub height, modern wind turbines are built at a height greater than 80 m, to capture more wind energy and lower cost per unit power output (European Wind Energy Association, 2009; Nelson, 2009).

3.2.WIND TURBINES

A wind turbine is a large rotating device that converts the energy of the wind into mechanical energy. This mechanical energy can then be used for simple mechanical tasks, such as turning large wheels to grind grain, in which case the wind turbine is referred to as a windmill. It can also be used as a source of electrical energy, either for a nearby application or after being transferred via high-voltage lines to a central switching station, in which case it is referred to as a wind generator (Figure 3.15).

The use of wind turbines as electrical generators has increased recently, becoming a major global source of power generation. Wind power is becoming more and more significant in the global transition away from fossil fuels and toward more sustainable energy sources. It is widely applicable, reasonably priced, and capable of being highly effective in a variety of global settings. While there are some environmental concerns regarding wind generators, these are usually small, and wind energy generates more than 120 gigawatts (GW) of electricity globally.

Horizontal-axis wind turbines (HAWT) and vertical-axis wind turbines (VAWT) are the two main categories into which turbines can be divided. Based on the well-known windmill design, HAWTs make up the majority of turbines that people are accustomed to seeing. Typically, modern HAWTs have three blades and are located at the summit of massive towers to reach higher altitudes where wind shear is significantly greater.

Figure 3.15. *The use of renewable energy sources such as wind energy for the generation of electricity is not a new technology.*

Source: By © Hans Hillewaert, CC BY-SA 4.0, https://commons.wikimedia.org/w/index.php?curid=6361901.

The majority of HAWTs are computer-controlled devices that automatically turn into the wind to maximize any available wind. With HAWTs, there is no need for any kind of backtracking because their blades are always traveling against the wind. Unlike VAWTs, therefore, they are producing power the entire time they rotate. Unfortunately, due to their size, HAWTs are very costly to transport; typically, transportation accounts for 20% of the total cost. They can also be visually unappealing due to their height, and installation is a challenging, expensive, and time-consuming procedure. In contrast, a VAWT has a rotator shaft that protrudes straight up into the air, and it is equipped with foils that capture wind energy to rotate the entire shaft.

The most prevalent variety is the Darrieus wind turbine, which resembles a big eggbeater in appearance. Compared to HAWTs, VAWTs can generate electricity at substantially slower wind speeds and require significantly less height. They are also significantly less expensive to transport and install. Nonetheless, they operate at about half the efficiency of HAWTs overall, and because they are lower to the ground, they frequently encounter much lower wind speeds. There are several non-standard designs of wind turbines in addition to the two main varieties, which account for the great majority of turbines worldwide. These are typically HAWT or VAWT derivatives, like the ducted rotor turbine or aerogenerator. Stranger variations have also been proposed and occasionally tested, like the airborne wind turbine and magnetic levitation wind turbine. However, the majority of alternative designs have not yet gained widespread commercial acceptance.

3.2.1. Wind Turbines Work

Wind energy is transformed into electricity by wind turbines. A wind turbine consists of two main parts: wind-driven blades and mechanical gears and an electricity generator. The rotor shaft rotates the gears inside the gearbox, just as the wind rotates the turbine blades. Energy is produced by this movement, which is then stored in the electricity generator and added to the power system. Wind turbines come in two varieties: vertical axis and horizontal axis.

The appearance of a HAWT is strikingly similar to that of a windmill. At the summit of the tower are the main rotor shaft and electrical generator. To automatically align wind turbines with the direction of the prevailing wind, servomotors, and wind sensors are frequently utilized. The large, slow rotation is changed into the rotational speed required to produce electricity by the gearbox (Figure 3.16).

Figure 3.16. *The blades of wind turbines are very stiff and made from a combination of hard plastics and light metals.*

Source: By Paul Anderson, CC BY-SA 2.0, https://commons.wikimedia.org/w/index.php?curid=5125761.

The blades are installed in front of the tower and are usually pointed upwind. This is done to prevent the blades from being forced back into the tower and to accommodate turbulence forces.

There are usually three blades on a wind turbine. They are between 65 to 130 feet (20 to 40 meters) long and lightly colored. In the event of high wind speeds, all wind turbines have a shutdown feature. This feature is used to avoid damage to the wind turbine equipment.

The HAWT and the VAWT are marginally different. The gearbox and generator in this model are situated at the turbine's base on the ground. This change eliminates the need to aim the turbine straight into the wind and offers more flexibility.

The mechanical components of the turbine are easier to access for maintenance and repairs because they are located at the bottom. Typically, VAWTs are mounted on a building's rooftop or the ground. At this altitude, the wind typically blows slower and thus produces less energy. Making a very long axis is one potential workaround, but it takes a lot of materials and resources.

Rooftop-mounted units pick up wind that has been redirected over the roof by the shape of the building. This feature can significantly increase the wind speed. A rooftop turbine should be installed so that the blades are at least half the building height off the roof. This is the ideal distance to capture the most wind energy (European Wind Energy Association, 2009; Nelson, 2009).

3.2.2. Wind Turbine Classification

Wind turbines can be classified according to the turbine generator configuration, airflow path relative to the turbine rotor, turbine capacity, the generator-driving pattern, the power supply mode, and the location of turbine installation.

3.2.2.1. Horizontal-Axis and Vertical-Axis Wind Turbines

Modern wind turbines can be divided into two categories based on how their rotor blades are arranged: horizontal-axis turbines and vertical-axis turbines. The horizontal-axis type, in which the blades rotate in a direction parallel to the wind stream, accounts for the majority of commercial wind turbines built today. This kind of wind turbine has low cut-in wind speeds, high power density, low cost per unit power output, and high turbine efficiency (Figure 3.17).

Figure 3.17. *The world's tallest vertical-axis wind turbine.*

Source: https://upload.wikimedia.org/wikipedia/commons/f/f9/%C3%89ole_%C3%A0_Cap-Chat_en_2010.JPG.

The vertical axes of the VAWT blades revolve in relation to one another, perpendicular to the ground. The ability of a VAWT to accept wind from any direction eliminates the need for yaw control, which is a major benefit. The wind tower's design and construction are significantly simplified, which lowers the turbine's cost because the wind generator, gearbox, and other major turbine components can be assembled on the ground. However, during initialization, the VAWT blades must rotate using an outside energy source. The wind turbine's maximum practicable height is consequently constrained because its axis is only supported at the ground on one end. Today, vertical-axis wind turbines only make up a small portion of wind turbines due to their lower wind power efficiency.

3.2.2.2. Upwind and Downwind Wind Turbines

HAWTs are further divided into upwind and downwind wind turbines based on how the wind rotor is arranged in relation to the direction of the wind. Upwind wind turbines, in which the wind rotors face the wind, make up the majority of HAWTs currently in use. Avoiding flow field distortion as the wind passes through the wind tower and nacelle is the primary benefit of upwind designs. Wind passes through the nacelle and tower of a downwind turbine before hitting the rotor blades. With this arrangement, the rotor blades can be made more flexible without taking tower strike into account. However, the wind power output produced by a downwind turbine varies significantly due to the influence of the distorted unstable wakes behind the tower and nacelle. Furthermore, an unstable flow field could lead to increased aerodynamic losses and fatigue loads on the turbine. Moreover, a downwind wind turbine's blades may make a louder, thumping, or impulsive noise (Figure 3.18).

Figure 3.18. *The downwind variable blades pitch control the wind turbine.*

Source: By Eileen at OE – Own work, CC BY-SA 4.0, https://commons.wikimedia.org/w/index.php?curid=68516070.

3.2.3. Wind Turbine Capacity

Based on their rated capacities, wind turbines can be categorized into several broad groups, including micro, small, medium, large, and ultra-large wind turbines. While there isn't a precise definition for micro wind turbines, it's generally agreed that a turbine with a rated power of less than a few kilowatts falls into this category. Micro wind turbines work particularly well in places without access to the electrical grid. They can be applied to specific structures, like water pumps, street lights, and inhabitants of isolated places, especially in developing nations. Micro wind turbines can be widely installed in most parts of the world to fully utilize wind resources and significantly increase the availability of wind power generation because they operate in moderate wind speeds and require relatively low cut-in speeds at start-up (Figure 3.19).

Figure 3.19. *Sweden's wind power capacity to surpass nuclear this year.*

Source: By VanGore – Own work, CC BY-SA 3.0, https://commons.wikimedia.org/w/index.php?curid=21243853.

When we talk about small wind turbines, we usually mean those that have an output power of less than 100 kW. Small wind turbines are widely used in homes, farms, and other isolated locations for things like telecom sites and water pumping stations in the countryside. Dispersed small wind turbines can boost local electricity production without necessitating an increase in transmission line capacity. With power ratings ranging from 100 kW to 1 MW, medium-sized wind turbines are the most widely used type.

These wind turbines can be used for distributed power, hybrid systems, village power, wind power plants, and other applications in both on- and off-grid settings. Large wind turbines are defined as megawatt wind turbines up to

10 MW. Multi-megawatt wind turbines have taken center stage in the global wind power market in recent years. Megawatt wind turbines are currently used in the majority of wind farms, particularly in offshore wind farms. Wind turbines larger than 10 MW are considered ultra-large wind turbines. We are still in the early phases of research and development for this kind of wind turbine.

3.2.3.1. Direct Drive and Geared Drive Wind Turbines

Wind turbines are categorized as direct drive or geared drive groups based on the state of the drivetrain in a wind generator system. A regular geared drive wind turbine usually uses a multi-stage gearbox to transfer the rotational speed from the low-speed shaft of the blade rotor to a fast rotation on the high-speed shaft of the generator rotor. This increases the generator rotor rotating speed and yields a higher power output. Geared generator systems have the advantages of being smaller in size and weight as well as less expensive. Nevertheless, the use of a gearbox can raise mechanical losses and turbine noise levels while drastically reducing wind turbine reliability (Figure 3.20).

Figure 3.20. *Wind turbine direct.*

Source: By Basotxerri – Own work, CC BY-SA 4.0, https://commons.wikimedia.org/w/index.php?curid=66533400.

By eliminating the multi-stage gearbox from a generator system, the generator shaft is directly connected to the blade rotor. Therefore, the direct-drive concept is more superior in terms of energy efficiency, reliability, and design simplicity.

3.2.3.2. On-Grid and Off-Grid Wind Turbines

Applications for wind turbines can be found both on and off the grid. Nearly all large- and medium-sized wind turbines are employed in grid-tied applications. The fact that there is no energy storage issue with on-grid wind turbine systems is one of their clear benefits.

On the other hand, the majority of small wind turbines are used off-grid for telecommunications, farms, and residential buildings. Wind power generated by off-grid wind turbines, however, is an erratic power source that can fluctuate significantly over brief periods of time with little notice. To increase the stability of the wind power supply, off-grid wind turbines are typically used in conjunction with batteries, diesel generators, and photovoltaic systems.

3.2.3.3. Onshore and Offshore Wind Turbines

Onshore wind turbines have a long history on its development. There are a number of advantages of onshore turbines, including lower cost of foundations, easier integration with the electrical-grid network, lower cost in tower building and turbine installation, and more convenient access for operation and maintenance (Table 3.2).

Table 3.2. *Summary of Literature Review on the Structural Integrity of Offshore Wind Turbines*

Component	Methodology	Materials/ Structure	Conditions	Key Findings
Blades	FEA	Fiberglass/ epoxy and carbon Fiber/epoxy composites	NACA4424, camber = 4–14%, thickness = 24%, IEC 61400–01 Class 1A loads	Stiffness of carbon fiber/epoxy blades is 3.7 times higher than fiberglass/epoxy blades. Maximum flap wise deflection is 2.9 m for critical wind load conditions.
		Composite	Pulse load and half-sine load	Geometric non-linearities increase flap deflection of the blade.

Component	Methodology	Materials/ Structure	Conditions	Key Findings
	FEA/Experimental	Leading edge protective layer made of thermoplastic polyurethane	Static and dynamic loading, 13 m long blade, 225 kW turbine	ANSYS and ABAQUS underestimate blade edgewise stiffness; Inclusion of blade protector increases the stiffness of the blade; At blade tip, von Mises, frictional, and tensile stresses are higher than the yield strength.
	GA	Material density = 1220–1920 kg/m^3, G = 2–12 GPa, E = 40–115 GPa	Blade thickness = 15–35 mm, 5 MW turbine, seismic conditions	Blade material anisotropy leads to higher flap wise and edgewise stiffness near the root
Sub-Structure	Experimental	Gravity-support	Excitation of an active mass damper	Damping ratios = 0.2 and 2.4% (first and second modes)
	GA	Gravity-support	Soil properties	Damping ratio is dependent on the soil stiffness
	FEA/GA	Monopile	Integrated optimization model, aero-hydro-elastic analyses, 5 MW turbine	Frequency = 0.2123 HZ (first mode); Deformation = 0.078 m; Mass saving = 19.8%
	FEA	Monopile	IEC 61400-01 Class 1A loads, 78 m high tower, 18 m long transition piece, 3 MW turbine	Maximum deformation, principal stress and elastic strain occur at the tower head; Monopile substructure avoids resonance effects

Component	Methodology	Materials/ Structure	Conditions	Key Findings
		Monopile	80 m high tower, 3.6 MW turbine, IEC Class 1A wind loading, wave loading, tidal loading, hydrostatic pressure	More stoppers in the transition piece are better for structural integrity but are costly; Modes 2 and 4 are affected by foundation scouring
		Tripod	6 MW turbine; tower height = 100 m, blades length = 80.35 m, wind, tidal, and fatigue loads	Foundation scouring has negligible effects on the natural frequency, deflection, and rotation of the tripod
		Jacket	Artificial Neural Network surrogate modelling, 10 MW turbine	Reliability index indicates turbine life of 18 years, instead of 22 years calculated theoretically
	Numerical/ Analytical	Mooring	200 m depth, catenary, and taut, clump weight and buoy, drag, and suction	Inclusion of clump weights and buoys are beneficial for catenary; catenary-taut hybrid mooring is more stable and cost effective
		Mooring	Tension leg, coupled dynamic response model	Most stable mooring has 4 legs, 12 cables and inclination angle = 30°
Tower	FSI	Monopile	Wind, wave, and seismic loads	Rayleigh's energy method is a close approximation of coupled FSI modelling

Component	Methodology	Materials/ Structure	Conditions	Key Findings
		Monopile	Concentrated spring model, 5 MW turbine	Natural frequencies predictive by FAST and ADAMS solver are in close agreement with SDOF approach
	FEA	Monopile	Siemens 2.3 MW turbine, P-y method,	First two bending modes occur at 0.35 Hz and 1.46 Hz
		Monopile	Realistic atmospheric stratifications 1.5 MW turbine and 63 m high tower	Rotor thrust significantly deflects the tower
		Rotor-hub-tower system	Wind wheel gravity, cabin whole gravity, wind tower whole gravity	Swing, bending, and torsion are the main vibration modes
	Experimental	Tower faults	FDD on a scaled model	Damage near the bottom of the tower is more impactful than at the top; the structural response is non-linear
	PSO	Jacket/lattice tubular	Loading on lattice and wind loads, 1.5 MW turbine	Natural frequency = 0.382 Hz; optimal columns' inclination angle = 85° and angle of diagonal members = 48.3°

Source: Asim T, Islam SZ, Hemmati A, Khalid MSU. A Review of Recent Advancements in Offshore Wind Turbine Technology. Energies. 2022; 15(2):579. https://doi.org/10.3390/en15020579.

Since the 1990s, offshore wind resources have outperformed onshore wind resources in terms of wind power intensity and continuity, which has led to a faster development of offshore wind turbines. When compared to an onshore turbine, an offshore wind turbine can produce more power and run for longer hours annually. Furthermore, compared to onshore locations, offshore locations have less stringent environmental regulations. For example, offshore wind turbine noise is no longer a problem.

3.3. CARBON FOOTPRINT IMPACT

Compared to conventional fossil fuels, wind energy has a lower carbon footprint, which makes it an important tool for lowering greenhouse gas emissions. Wind turbines work by harnessing the kinetic energy of the wind to create power without releasing carbon dioxide into the atmosphere. The production, shipping, installation, and upkeep of wind turbines are the main sources of carbon emissions related to wind energy.

When compared to the lifetime emissions of fossil fuel-based power generation, these emissions are, nevertheless, comparatively low. Over the course of a turbine's lifetime, wind energy typically emits 10 to 20 grams of CO_2 equivalent emissions per kilowatt-hour (g CO_2e/kWh). This is a substantial reduction from the emissions produced by coal (800–1,000 g CO_2e/kWh) or natural gas (450–550 g CO_2e/kWh) power. The lack of continuous carbon output during the generation of electricity significantly offsets the emissions from wind energy throughout a wind turbine's operational life, which can exceed 20 years. As a result, wind energy efficiently lowers carbon emissions, helps mitigate the effects of climate change, and facilitates the switch to a low-carbon, more sustainable energy system (European Wind Energy Association, 2009; Nelson, 2009).

3.3.1. Analysis of Wind Energy's Contribution to Emission Reductions

Since wind energy replaces fossil fuel-based power generation with a low-carbon alternative, it significantly reduces emissions. By using the kinetic energy of the wind, wind turbines generate electricity without directly emitting greenhouse gases into the atmosphere. The main emissions related to wind energy are produced during the wind turbines' construction, transportation, installation, and maintenance stages.

The total carbon footprint of wind energy is significantly smaller than that of conventional fossil fuels, even with these lifecycle emissions. According to lifecycle assessments, wind energy produces between 10 and 20 grams of

CO_2-equivalent emissions for every kilowatt-hour (g CO_2e/kWh) of electricity produced. On the other hand, natural gas plants generate roughly 450 to 550 g CO_2e/kWh, whereas coal-fired power plants release between 800 and 1000 g CO_2e/kWh. The lack of direct carbon emissions during the generation of electricity effectively offsets the emissions from wind energy many times over during the operational life of a wind turbine, which frequently exceeds 20 years.

Moreover, wind energy directly lowers the amount of carbon dioxide and other pollutants released into the atmosphere as it replaces fossil fuel generation. To meet climate targets and mitigate global warming, this displacement is essential. Countries can significantly reduce their greenhouse gas emissions, improve energy security, and make the transition to a low-carbon and more sustainable energy system by increasing the deployment of wind energy.

3.3.2. Challenges and Solutions: Intermittency, Land Use, and Integration into Power Grids

Wind energy, while a crucial component of a low-carbon energy future, faces challenges related to intermittency, land use, and integration into power grids.

Addressing these challenges requires targeted solutions to maximize wind energy's benefits:

1. **Intermittency** refers to the variable nature of wind power, as wind speeds fluctuate and do not always align with electricity demand. This can lead to periods of insufficient power generation. Solutions include enhancing energy storage technologies, such as batteries and pumped hydro storage, to store excess energy generated during high-wind periods for use during low-wind periods. Additionally, implementing demand response strategies can help balance supply and demand by adjusting energy consumption patterns.
2. **Land Use** concerns arise from the space required for wind farms, which can impact local ecosystems and land availability. To mitigate this, wind farms can be sited in less ecologically sensitive areas and on lands with limited alternative uses, such as agricultural lands or offshore locations. Advancements in turbine design, such as taller towers and larger blades, can also increase energy output per unit of land.
3. **Integration into Power Grids** involves incorporating wind energy into existing power systems, which can be challenging due to its variability and the need for grid stability. Solutions include developing advanced grid management technologies and infrastructure improvements, such as smart grids and enhanced

transmission networks, to better accommodate and distribute wind energy. Investing in grid flexibility and resilience through decentralized energy systems and improved forecasting tools can also facilitate smoother integration.

Addressing these challenges with innovative solutions ensures that wind energy can effectively contribute to a cleaner, more sustainable energy mix while supporting grid stability and environmental preservation (European Wind Energy Association, 2009; Nelson, 2009).

REFERENCES

1. Brandt, K., & Zeumer, J. (2007). *Wind Power Plant Comprising a Rotor Blade Adjusting Device* (U.S. Patent No. 7,256,509).
2. Cabezón, D., Martí, I., Isidro, M. J. S., & Pérez, I. (2004). Comparison of methods for power curve modelling. *Global Windpower 2004*, Chicago, Illinois.
3. El-Ali, A., Moubayed, N., & Outbib, R. (2007). Comparison between solar and wind energy in Lebanon. In *Proceedings of the 9th International Conference on Electrical Power Quality and Utilisation*, Barcelona.
4. European Wind Energy Association. (2009). *Wind Energy–The Facts, Part I: Technology, The Annual Variability of Wind Speed.*
5. Kaźmiekowski, M. P., Krishnan, R., & Blaabjerg, F. (2002). *Control in Power Electronics*. Academic Press.
6. Ko, K. N., Kim, K. B., Kang, M. J., Oh, H. S., & Huh, J. C. (2009). Variation in wind characteristics for 30 years on Jeju Island, Korea. In *Proceedings of the 2009 European Wind Energy Conference*, Marseille, March.
7. Nelson, V. (2009). *Wind Energy–Renewable Energy and the Environment*. CRC Press.
8. Noguchi, T. (2006). *Windmill for Wind Power Generation* (U.S. Patent No. 7,084,523).
9. Paulsen, B. M., & Schroeder, J. L. (2005). An examination of tropical and extratropical gust factors and the associated wind speed histograms. *Journal of Applied Meteorology, 44*(2), 270–280.
10. Ulgen, K., & Hepbasli, A. (2002). Determination of Weibull parameters for wind energy analysis of Izmir, Turkey. *International Journal of Energy Research, 26*(6), 495–506.
11. Zervos, A. (2009). Wind power as a mainstream energy source. In *Proceedings of the 2009 European Wind Energy Conference*, Marseille, March.

Chapter **4**

Hydropower

Contents

INTRODUCTION

Although hydropower is frequently referred to as a renewable resource, it is actually best understood as a separate entity due to its many differences from other renewable resources. With costs comparable to those of the main energy sources, it is a proven source that currently provides around 3% of the world's energy production. The cost of renewable energy comes from the fact that, despite its abundance, it must be gathered from vast regions. When it comes to hydropower, the river valleys do the collecting for us. The rivers are dammed to create lakes to guarantee a constant flow of water and to deepen the fall. Electricity is generated by the water passing through the turbines. A large portion of the energy in mountainous nations like Austria, Switzerland, Iceland, and Norway comes from hydroelectric power.

Unlike other renewable energy sources, hydropower is characterized by its high degree of dependability and inherent risk associated with the potential for dam failures. Since most of the suitable rivers in the developed world are already in use, hydropower cannot be greatly expanded. There are several potential locations in developing nations, but they are typically remote from population centers and thus open to guerrilla attacks. The pylons that carry electricity from the CalboroBassa dam on the Zambesi river have frequently been blown up by saboteurs. Large tracts of land are needed for hydropower, and eventually, the lakes get silted up and become unusable. Several nations that rely heavily on hydropower—Norway, New Zealand, Brazil, and the USA—have encountered severe issues during protracted droughts.

Large tracts of land are invariably submerged by hydropower, frequently including villages and productive agricultural land. Additionally, it may upset the nearby ecosystems. Fish ladders can be built to get around the dam, which stops fishlike salmon from migrating upstream. One significant drawback of the Aswan dam on the Nile is that it deprives the areas beneath it of silt, which is beneficial to the soil. There are a few locations where submerged land can produce electricity hydroelectrically by allowing seawater to flow into it. It has been calculated that the Quattara depression in North Africa could produce 4000 MW in this manner.

4.1. OVERVIEW OF HYDROPOWER SYSTEMS

One of the earliest and most significant forms of renewable energy is hydropower, also known as hydroelectric power, which produces electricity by harnessing the natural flow of flowing water. At present, hydropower makes up 37% of all U.S. renewable energy production and roughly 7% of all U.S. production of electricity. Hydropower facilities come in various sizes, although the Hoover

Dam, a massive structure that harnesses the power of an entire river behind its wall, may be the most commonly recognized example of this energy source. They can be small or very large, depending on the water flows in irrigation ditches or municipal water systems. With run-of-river features or diversions that direct a portion of a stream through a power plant before the water returns to the main river, they can even be considered "damless." Regardless of the technique, most people are unaware of how much more accessible and widely used hydropower is. In actuality, hydropower is used to generate electricity in every state but Delaware and Mississippi, albeit to varying degrees. For instance, hydropower accounts for roughly 74% of the electricity in the state of Washington.

Water moving through the hydrological cycle—which is fueled by solar radiation—is used to produce hydropower. Where water is present, incoming solar radiation is absorbed at the surface of the land or sea, heating the surface and causing evaporation. Nearly half of all solar radiation that reaches Earth's surface is used to evaporate water, which in turn powers the hydrological cycle. This cycle therefore has enormous potential energy, but only a very small amount could be technically realized. The amount of water vapor in the air rises as a result of evaporated water entering the atmosphere. Air and its vapor content are moved over the surface of the Earth by global, regional, and local wind systems, which are created and sustained by temporal and spatial variations in solar energy input. These wind systems can travel thousands of kilometers from the site of evaporation. Eventually, roughly 78% of the vapor condenses and falls as precipitation on land and 22% on oceans. As a result, there is a net movement of water from the oceans to the Earth's surface and a correspondingly significant flow of water—river and groundwater runoff—back to the oceans. Hydropower can be produced from the flow of water in rivers, or more accurately, from the energy of water traveling by gravity from higher to lower elevations on its return trip to the ocean.

4.1.1. What Is Hydropower?

Hydropower, sometimes referred to as water power, is the process of using falling or swiftly flowing water to generate electricity or run machinery. This is accomplished by harnessing a water source's kinetic energy or gravitational potential to generate electricity. One way to produce energy sustainably is through hydropower. Nowadays, hydropower is mostly utilized to generate hydroelectric power. It is also utilized as one half of pumped-storage hydroelectricity, an energy storage system. Since hydropower produces no carbon dioxide or other air pollutants directly and offers a reasonably reliable power source, it is a desirable substitute for fossil fuels. However, it has drawbacks in the areas of economics, society, and the environment. It also needs a water source that is sufficiently energetic, like a river or an elevated lake. The World Bank and other international organizations consider hydropower to be a low-carbon economic development tool.

4.1.2. History of Hydropower Development

Before commercial electric power became widely available, hydropower was used for irrigation and to run a variety of machines, including sawmills, watermills, and textile mills. Humans have improved their way of life by cooperating with nature to generate electricity through the use of water. An ancient resource still in use for services and beneficial purposes is the mechanical power of falling water. More than 2,000 years ago, the Greeks used it to power water wheels that ground wheat into flour. Mechanical hydropower was widely employed in the 1700s for pumping and milling. The advancement of water turbines persisted in the 1700s and 1800s. In Cragside, Rothbury, England, the first hydroelectric power plant was set up in 1870. In Grand Rapids, Michigan, a dynamo powered by a water turbine was employed to light theaters and storefronts in 1880, marking the beginning of the industrial use of hydropower. In 1881, street lighting was established in Niagara Falls, New York, thanks to a brush dynamo that was linked to a flour mill's turbine. The world's first hydroelectric station (with a capacity of 12,5 kW) was put into service on September 30, 1882, at the Vulcan Street Plant in Appleton, Wisconsin, USA, on the Fox River. This was the turning point because it allowed the electric generator to be connected to the turbine, lighting a home and two paper mills.

The fossil fuel-fired plants of the past were far less dependable and efficient than the early hydropower plants. As a result, small- to medium-sized hydropower plants proliferated everywhere there was a sufficient supply of flowing water and a demand for electricity. The quantity and size of hydropower, nuclear, and fossil fuel plants expanded along with the electricity demand. Concurrently, worries about the effects on society and the environment surfaced.

Hydropower plants (HPP) today span a very large range of scales, from a few watts to several GW. The largest projects, Itaipu in Brazil with 14,000 MW[2] and Three Gorges in China with 22,400 MW[3], both produce between 80 to 100 TWh/yr (288 to 360 PJ/yr). Hydropower projects are always site-specific and thus designed according to the river system they inhabit. Historical regional hydropower generation from 1965 to 2009 is shown in Figure 4.1.

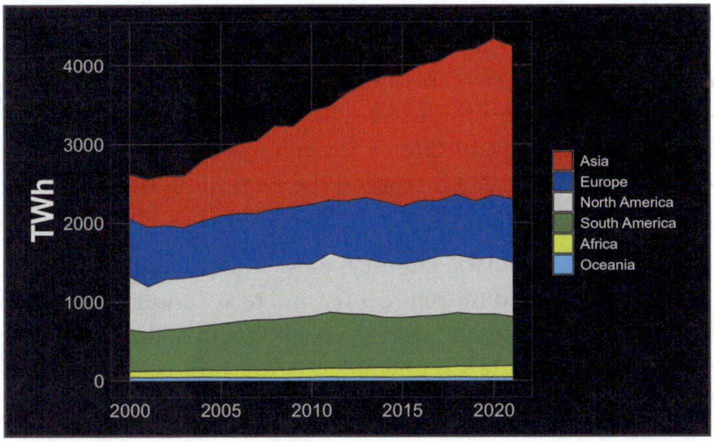

Figure 4.1. *Hydropower generation (TWh) by region.*

Source: By Wizmut – Own work, CC0, https://commons.wikimedia.org/w/index.php?curid=143229752.

The technology can meet large centralized urban energy needs as well as decentralized rural needs because of the wide range of sizes of hydropower plants. HPP operate in isolation and supply independent systems, frequently in rural and remote parts of the world, even though their main contribution to the current global energy supply is the generation of electricity as part of centralized energy networks. Hydro energy can also be used to heat and cool spaces or to satisfy mechanical energy requirements. If there is a local objective to use hydrogen as fuel for transportation and hydropower is abundant in a region, hydroelectricity has also been studied more recently for use in the electrolysis process for producing hydrogen fuel.

The water used in HPP does not power the turbines. After producing electricity, the water can be used for several other vital purposes. Actually, a sizable percentage of hydropower projects are made with multiple uses in mind. In these cases, the dams can improve conditions for fishing, tourism, leisure, and navigation. They can also help prevent or lessen floods and droughts, offer the ability to irrigate agricultural land, and supply water for home, municipal, and industrial use. When discussing hydropower and the various uses of water, one point that is frequently missed is that the power plant, being a source of income, can occasionally assist in covering the costs of the infrastructure needed to develop other water uses that might not produce enough direct revenue to finance their construction.

4.1.2.1. What Is the History of Hydropower?

Hydropower has thousands of years of history. For instance, over 2000 years ago, the Greeks ground wheat into flour using water wheels. The mid-1700s saw the writing of Architecture Hydraulique by French hydraulic and military engineer Bernard Forest de Bélidor, which marked the beginning of the development of the modern hydropower turbine.

The first half of the 19th century saw many significant advances in hydropower technology, and more recently, the past century has seen a number of hydroelectric advancements that have helped hydropower become an essential component of the US renewable energy mix (California Department of Water Resources, 2005; Towler, 2014).

4.1.3. How Does Hydropower Work?

Water flowing in on one side and out on the other, far below, is used by hydropower technologies to create electricity by taking advantage of the elevation difference produced by a dam or diversion structure. The Water Power Technologies Office (WPTO)'s research and development endeavors in the field of hydropower are showcased in the Department of Energy's Hydropower 101 video, which also explains how hydropower operates.

4.1.3.1. What Is the Cost of Hydropower?

Hydropower is a less expensive source of electricity than most others. States that primarily rely on hydropower, such as Idaho, Washington, and Oregon, have lower energy costs than the national average because hydropower only uses the energy from moving water. Hydropower also has relatively low maintenance, operating, and fuel costs over the course of a project's entire lifetime when compared to other electricity sources.

Like any significant energy source, hydropower will inevitably incur significant upfront costs; however, these costs will be dispersed over time due to its longer lifespan. HPP also typically use equipment that lasts longer between maintenance and replacement, which results in long-term cost savings.

Large hydropower facilities' installation costs are primarily made up of costs associated with electromechanical equipment (the machinery that generates electricity) and civil construction projects (such as the construction of dams, tunnels, and other essential infrastructure). Since hydropower is a site-specific technology, by carefully choosing the location and design, these costs can be reduced during the planning phase.

4.1.3.2. The Benefits of Hydropower

For thousands of years, people have understood and used the advantages of hydropower. HPP are not only an economical and environmentally friendly energy source, but they can also instantly supply electricity to the grid, making them a flexible and dependable backup source in the event of significant power outages or disruptions. In addition to producing energy, hydropower also provides many other advantages like flood control, irrigation support, and water supply.

4.1.3.3. How Do We Get Energy from Water?

A dam or other diversion structure is used to change the natural flow of a river or other body of water to produce electricity, a renewable energy source known as hydropower or hydroelectric power. Hydropower uses water as its fuel, which is neither diminished nor eliminated during the production of electricity and relies on the infinite, perpetually replenishing system of the water cycle. HPP come in a variety of forms, but they are all driven by the kinetic energy of water moving downstream. That kinetic energy is transformed into electricity by hydropower using turbines and generators, which is then fed into the electrical grid to power residences, companies, and industries.

4.1.3.4. How Exactly Is Electricity Generated at Hydropower Plants?

Plants that use hydropower are typically found on or close to a water source because the technology uses water to create electricity. The amount of energy that the flowing water can produce is determined by the head, or change in elevation, as well as the volume of water flow. The amount of electricity that can be produced increases with flow and head height. Water travels through a pipe, also called a penstock, at the plant level, spinning a turbine's blades, which in turn spins a generator, which generates electricity. This is how the majority of conventional hydroelectric facilities, such as pumped storage and run-of-the-river systems, function (California Department of Water Resources, 2005; Towler, 2014).

4.2. HYDROPOWER TECHNOLOGY DEVELOPMENT

The largest renewable electricity source in the U.S. is hydroelectric power, which over the past 10 years has produced roughly 63.3% of the total electricity produced in the country. There are still plenty of opportunities to increase the country's hydropower resources through non-powered dams, water conveyance systems, pumped storage hydropower, and new site development, even after a century of demonstrated experience with this dependable renewable resource. Through the development and implementation of new technologies and essential components, as well as by identifying critical opportunity areas through which hydropower generation can be enhanced, the Water Power Program supports the hydropower industry and enhances current investments.

Pumped-storage hydropower, which serves as utility-scale grid storage technology, is a crucial component of DOE's renewable energy portfolio in addition to conventional hydropower. To demonstrate the advantages of pumped-storage hydropower as a component of our clean energy future, the Water Power Program can be very helpful. It can serve as a renewable means of grid stabilization and a catalyst for the high penetration of variable renewable energy sources (like wind and solar). The 2015 report from the Department of Energy to Congress lists important initiatives that could hasten the development of pumped storage in the U.S.

With more than 2500 U.S. companies supporting the hydropower industry, adding additional hydropower generation will create a large and enduring economic benefit by revitalizing the domestic manufacturing and hydropower industry.

The Water Power Program's work in the following areas of hydropower technology development:

- Low-Head Hydropower.
- Materials and Manufacturing.
- Hydropower Systems.
- Hydropower Technology Accomplishments.
- Pumped-Storage Hydropower.

1. **Low-Head Hydropower:** Across the nation, there is a huge chance to develop new low-head hydropower generating capabilities (i.e., those that function in an elevation range of two to 20 meters). These kinds of waterways can be found in many different parts of the United States at currently existing non-powered dams, canals, and conduits. The Water Power Program is funding the development of cutting-edge low-head hydropower technology, such as the first Archimedes Hydrodynamic Screw system installed in the country by Percheron Power. This project proved the low-head technology's simplicity, dependability, and affordability.

2. **Materials and Manufacturing:** The Water Power Program provides funding to R&D to find and test novel materials and production processes that will enhance hydropower's efficiency and cut costs. Research supported by the program focuses on coatings or materials that lower the lifetime costs of draft tubes, penstocks, and turbine runners. Finding and evaluating strategies to raise generator dependability and efficiency is another major goal of R&D.
3. **Hydropower Systems:** To lower operating and maintenance costs, increase unit availability and plant capacity factors, lower risk through improved system reliability, and improve the quality of the energy produced—both in terms of its environmental performance attributes and ancillary power benefits—the Water Power Program develops and tests new technologies and techniques. The optimization of water use, the use of cutting-edge materials and production techniques, and the evaluation of the worth of water power grid services are among the areas of emphasis. For instance, the United States' current hydropower facilities are beginning to deteriorate, and the information utilized to assess them is dispersed and out of date. Working with partners, the Water Power Program is integrating and updating data to comprehend the reductions in capacity factors, facility availability, and electricity generation.
4. **Technology Development Accomplishments:** The program has numerous accomplishments in hydropower technology development. The projects described below highlight just a few of the program's new opportunities and recent successes in cost reductions, water-use optimization, and facility upgrades.
5. **New Opportunities for Advanced Hydropower R&D:** The Water Power Program issued its first significant call for hydropower research and development (R&D) in over 10 years in 2011, following a reorganization of its hydropower technology efforts. These projects seek to lower the cost of hydropower technologies while showcasing the advantages of pumped storage and advanced hydropower for a dynamic grid. For instance, Natel Energy created, assembled, and tested a dependable powertrain for the Schneider Linear hydro-Engine in collaboration with Alden Research Laboratory. This powertrain can create new low-head hydropower capacity by lowering capital and operating expenses, resulting in a levelized cost of energy savings of about $2 per megawatt hour.

4.2.1. Optimizing Hydropower Systems for Power and Environment

A group of U.S national laboratories was sponsored by the Water Power Program to create and present the Water-Use Optimization Toolset (WUOT), a sophisticated collection of integrated analytical tools. Enhancing the environmental benefits of improved hydropower operations and planning, WUOT helps managers and operators run their HPP more efficiently, producing more energy and grid services from available water resources. WUOT encompasses tools for environmental performance operations, day-ahead scheduling, seasonal hydro-systems analysis, hydrologic forecasting, and real-time operations.

The following locations are deploying WUOT for demonstration:

1. The Oroville Complex on the Feather River in California.
2. The upper Colorado River portion of the Colorado River Storage Project.
3. The Conowingo Dam on the Susquehanna River in Maryland.

4.2.2. Revitalizing American Infrastructure

Sponsored through the American Recovery and Reinvestment Act of 2009, the Water Power Program completed three hydropower efficiency projects with overwhelming success—resulting in an increase of more than 3000 megawatt-hours per year.

1. The Los Alamos County Department of Public Utilities installed a low-flow turbine in its Abiquiu Hydroelectric Facility in New Mexico. The new turbine boosts overall facility output from 13.8 megawatts to 16.8 megawatts.
2. The City of Boulder in Colorado completed a modernization project to its Boulder Canyon Hydroelectric Project by installing a new turbine/generator unit. The new unit resulted in a 30% increase in generation and an 18%–48% increase in turbine efficiency.
3. The City of Tacoma installed two Francis turbine/generator units at the Cushman Dam in Washington. The new units add approximately 3.6 megawatts of annual electrical generation.

4.2.3. Hydropower and Other Water Energy Technologies

Water technologies are a broad category of systems that generate thermal or electrical energy from freshwater or the ocean. The most well-known application of water technology is hydropower, in which the flow of water powers a turbine that generates electricity through a generator. Since the water cycle naturally replenishes the fuel for hydropower and other water technologies, they are clean alternatives to burning fossil fuels that contribute to climate change. Dissimilar to other fuel-burning plants, hydropower does not require the purchase of fuels for generation. The building and maintenance of the generating facilities are the only expenses.

Hydropower generates roughly 15% of the electricity produced worldwide. 2014 saw a capacity of roughly 101,000 megawatts (MW), HPP provided 48% of the nation's renewable electricity and 6% of the nation's total energy. While the United States and other countries have developed the most suitable sites for large-scale dams, there are still plenty of opportunities to install hydropower systems at existing dams that are currently incapable of producing electricity, as well as to use other water energy technologies in rivers, tidal zones, and open ocean. In two 2012 studies conducted by the U.S. Department of Energy, new installations (including those that harness waves and tidal currents) could potentially provide hydropower 15% of America's electricity by 2030 (vs. currently non-producing dams, which could provide 12,000 MW of additional capacity: 6% as of right now). Hydropower facilities can be installed on rivers, oceans or lakes (California Department of Water Resources, 2005; Towler, 2014).

4.2.4. Rivers

4.2.4.1. Dams

The most advanced sources of water energy are large hydropower dams built on significant rivers. To release water for use when the river is running more slowly or during periods of peak energy demand, pumped storage or reservoir plants store water in a reservoir. This makes base-load power generation dependable. These massive structures include the Grand Coulee Dam in Washington and the Hoover Dam in Nevada. Large dams also provide for a variety of societal needs, including recreation, flood control, and irrigation.

The disadvantages of reservoir plants are numerous. Large reservoirs in tropical and boreal regions may release as much greenhouse gas into the atmosphere as a fossil fuel power plant, according to studies. As flooded vegetation breaks down, methane, and carbon dioxide are released in large amounts at the beginning of a dam's life and in smaller amounts as the dam is used. Additional effects include shifting the temperature of the water, releasing dissolved oxygen and other nutrients, damaging the ecosystem of the river, uprooting communities due to changes in the river's flow, and creating instability in the riverbanks that causes erosion, flooding, and deforestation. A vulnerability to climate change is hydropower. Long-term droughts can reduce a river's water level, which reduces the amount of electricity generated; on the other hand, glaciers melting quickly, snowpack melting quickly, or shifts in precipitation patterns from snow to rain can drastically alter a river's flow.

4.2.4.2. Run-of-the-River

Although run-of-the-river plants do not have water storage facilities, they may use low-level dams to raise the water intake level and turbine differential. In this instance, the power produced varies based on the river's cycle and is produced by the natural flow of the river. Run-of-the-river technology is typically used to provide less than 30 MW of electricity to individual communities, though it can also be utilized for large-scale power generation. Although this method of generating power is well-liked in China's rural areas, it has potential use across many regions, including the U.S. Compared to large hydropower dams, run-of-the-river technology usually disrupts much less of the river flow.

4.2.4.3. Current

Underwater, current generation functions similarly to a wind turbine. A given speed of moving water will produce a lot more power than a similar speed of wind because water is denser than air. However, the turbine itself needs to be more robust, which raises the cost. It's unclear how the current turbines will affect the environment. Fishermen's populations may suffer, but fish-safe turbines are already in existence.

There are numerous possible locations for the current generation in the United States, and several projects, such as those in the San Francisco Bay and the East River in New York, are currently in progress. First published by the Federal Energy Regulatory Commission, the U.S. license for a commercial tidal energy pilot project in 2012. The East River (Roosevelt Island Tidal Energy) project is moving toward constructing 30 turbines to produce 1 MW thanks to the 10-year license (California Department of Water Resources, 2005; Towler, 2014).

4.2.5. Oceans

4.2.5.1. Tidal Barrage

Utilizing the tides' predictable energy cycle, ocean tidal power generates energy. A tidal barrage is positioned at the mouth of a bay or estuary and functions similarly to a sizable hydropower reservoir dam. Power is produced by releasing the bay's retained water through the barrage's turbines. For the barrage to operate profitably, the tidal range between high and low tide must be sufficiently wide, approximately 10 feet. The most promising locations are in northern Europe and the U.S. Coast of the West. With a 240 MW capacity, the La Rance tidal barrage in France has been in operation since 1967. Because barrages are constructed in fragile estuary ecosystems, their potential environmental impact may be substantial; however, less invasive designs, like fences or floating barges, are being developed.

4.2.5.2. Tidal Current

Turbines anchored to the ocean floor or suspended from a buoy in the path of an ocean current could be used to generate power, much like river current technologies. Even though this technology is still in the research and development phase, the Gulf of Maine, North Carolina, the Pacific Northwest, and the Gulf Stream off Florida are among the possible US locations.

4.2.5.3. Wave

Waves are produced when wind blows across the ocean's surface, imparting energy to the water. Waves are predictable and always form, despite their fluctuating size and speed. Within the U.S., there are 2,100 terawatt hours of wave energy produced annually in coastal waters alone.

Many technologies are being tested to generate electricity from wave energy. Most systems use pressure differences slightly below the surface or harvest energy from waves at their surface. These systems work by applying pressure to the swells of the waves, which drives hydraulic pumps or pressurized air, which turns on generators. Although the effects of wave generators on the environment are not entirely understood, they are believed to be small and site-specific.

Strong wind currents near the ocean are ideal conditions for wave generation. These regions are found on the west coasts of continents, close to the equatorial trade winds, in polar regions with regular storms, and between 30° and 60° latitude. For offshore energy farms, hybrid wind and wave technology is being developed. The East Coast and the Pacific Northwest are two possible locations for hybrid wind-wave energy farms in the United States.

4.2.5.4. Ocean Thermal Energy Conversion

Steam from warm surface water is used in ocean thermal energy conversion (OTEC) to spin turbines. Steam is recycled into water by condensing it back into cold, deep ocean water. The temperature differential between the deep and surface waters must be 36°F. Tropical islands are among the possible locations. Because deep water pumping to surface generating stations is expensive, offshore wind energy is still in its early stages of development and is not yet economically viable. Ocean thermal air conditioning systems can be used with OTEC. Moreover, aquaculture can benefit from the deep water's abundance of nutrients. Plankton and algae can be grown in surface ponds that are pumped with deep water, in addition to salmon, lobster, and other seafood.

4.2.5.5. Ocean/Lake Thermal Air-Conditioning

Water has another use besides producing electricity: it can produce direct thermal energy. Buildings can have air conditioning thanks to water from lakes or oceans. By using cold, deep water to cool fresh water that runs through a building in a closed-pipe system, air conditioning can be achieved at a lower cost than with more conventional methods. To restart the cycle, the spent water is dumped back into the lake or ocean. For economic reasons, the cold deep water needs to be close to shore and between 39°F and 45°F. Examples of ocean thermal cooling systems are found in Toronto, where downtown buildings are air-conditioned using water from Lake Ontario, and Hawaii, where OTEC facilities are co-located. Large-scale OTEC projects (100 MW+) located in island nations like Guam, Hawaii, or Puerto Rico may be financially feasible.

4.2.6. Six Emerging Hydropower Technology Trends

These technologies cover hydropower flexibility, digitalization, storage, and variable speed turbines, generators with current-controlled rotors, as well as novel small-scale and fish-friendly technologies.

4.2.6.1. Hydropower Flexibility

Strategies for mitigating climate change must use a large percentage of renewable energy sources (RES). Modern hydropower technologies must offer more flexibility across a wider range of hydraulic conditions to support the variable generation of RES. A few of the most notable flexibility technologies currently in development are two-phase air-water injection along the axis; ejector power plants for the excess flow rate; hydroelectric energy storage solutions; J-grooves; air injection/admission; axial water injection with high/low velocity and low/high discharge; tangential water injection at a cone wall; axial water injection with a counter-flow tangential component; stabilizer fins; and smart modeling and control solutions to increase unit flexibility and operating range.

4.2.6.2. Hydropower Digitalization

Advanced grid supporting services can be provided by gathering and analyzing real-time data to modify the operational conditions of hydropower turbines without endangering the dependability and security of the stations. Hydropower digitalization is predicted to increase current hydropower energy production by 42 TWh in total. A substantial decrease in greenhouse gas emissions and an annual operational savings of USD 5 billion could result from such an increase.

4.2.6.3. Energy Storage and Variable Speed Turbines

Packed hydropower storage (PHS) is one type of energy storage that power systems need to counteract the rise in variable RES. Fixed or variable speed turbines can improve peaking regulation in PHS plants, which operate in both pump and turbine mode. With the use of electronic converters, speed can be changed in one of two ways: converter-fed synchronous machines, that is, synchronous machines with a variable-frequency driven stator; and double-fed induction machines, that is, electric devices in which the rotor and stator windings are supplied with AC currents. The first large variable-speed hydropower plant in Europe, the Goldisthal PHS plant in Germany, was put into service in 2004. It has four 265 MW pumped-storage units, two of which are connected to the grid by a doubly-fed induction machine.

4.2.6.4. Generators with Current-Controlled Rotors

Better control of electrical machines can be ensured by novel power electronics with current-controlled power supplies, for instance during the start and stop phases. For instance, the hydropower plant Frades II in Portugal has two pump turbines, and the induction motor generator's rotor power is managed by an AC-excitation system.

4.2.6.5. Novel Small-Scale Hydropower Technologies

A key component of mini-grid and rural electrification initiatives is small-scale hydropower. Specifically, new designs for turbines (such as very low head turbines and hydrokinetic turbines) and gravity hydraulic machines (such as water wheels and Archimedes screws) as well as advanced designs and operation strategies for pumps acting as turbines (PATs) are being introduced. PATs are typical pumps with turbine mode capability. The Deriaz turbine is one of the additional designs being developed for the PAT context; however, this particular turbine is typically found in large hydropower plants.

4.2.6.6. Fish-Friendly Hydropower Technologies

Hydropower has raised environmental concerns, which is why plant sustainability is being taken into account more and more during the design phase. Archimedes hydrodynamic screws, water wheels, and Vortex turbines for low head applications, as well as the Alden turbine and the minimum gap runner turbine for higher heads, are examples of modern fish-friendly turbines.

4.2.7. Future of Hydroelectric Power

There are several advantages to hydropower generation as well as some disadvantages. Technological developments in hydropower generation will lead to fewer issues and better/more effective results. One way to improve efficiency and recover losses caused by drought is to update the current hydropower infrastructure and make sure the plants can continue to run for many years to come. Upgrading outdated plants will cost $127 billion over the course of the next 10 years. This represents almost 25% of the entire hydropower investment made worldwide and nearly 90% of the investment made in North America and Europe. Investments in new plants are being made in nations other than the USA. Over 75% of hydropower capacity through 2030 will come from large projects in Asia and Africa, but some are concerned about the effects on the environment.

Because run-of-river facilities don't have reservoirs, they have less of an environmental impact, but they can't produce energy on demand because their output is dependent on seasonal flow. This decade, run-of-river hydropower is projected to contribute 13% of capacity additions overall, whereas pumped hydro will contribute 29% and traditional hydropower 56%. However, the growth of hydropower is generally slowing and is expected to decrease by 23% by 2030. Reversing this trend will require tight regulations and permitting procedures, as well as the establishment of strict sustainability guidelines and emissions monitoring plans to win over the public. Developers would find it easier to secure power purchase agreements with a shorter development timeline, which would encourage investment because returns would be assured. Weaning the world off fossil fuels will require finding the correct incentives for hydropower and pumped storage development, and doing so sustainably.

4.2.8. Technological Aspects

Hydropower is a well-established technology for generating electricity from the energy of flowing or falling water. Here's a rundown of the key technological aspects involved:

4.2.8.1. System Aspects

4.2.8.1.1. Photovoltaic System Components

To make use of the electricity from photovoltaic cells and modules, one has to build a complete system, also comprising electronic parts, support structures, and sometimes electricity storage. It is customary to use the term balance-of-system (BOS) for the sum of system components and installation excluding modules.

4.2.8.1.2. Type and Size of Photovoltaic Systems

Photovoltaics has a broad range of applications, ranging from grid-connected rooftop systems and large power plants to consumer goods and small standalone units for rural use (like solar home systems and solar lanterns). Systems with battery storage typically range in size from 50 watts to 1 kilowatt; rooftop grid-connected systems range in size from 500 watts to 5 kilowatts; grid-connected ground-based systems and larger building-integrated systems range in size from 10 kilowatts to many megawatts. Out of these market segments, focused marketing efforts and financial incentives are expected to drive the fastest growth in rural electrification for sustainable development and building-integrated systems (as precursors to large-scale implementation).

4.2.8.1.3. Need for Storage

Since photovoltaic modules provide an erratic energy source, the majority of standalone systems have battery storage (often a lead-acid battery) to supply power at night or on cloudy days. Batteries can occasionally hold energy for extended periods of time. Grid-connected photovoltaic systems allow the grid to act as "virtual storage," reducing the amount of fuel used by power plants that burn coal, oil, or gas by the amount of electricity supplied into the grid from photovoltaics.

4.2.8.1.4. Performance Ratio of Photovoltaic Systems

Being able to forecast how much energy a solar system of a given size will actually contribute to the grid is very important from a practical standpoint. However, to do that, accurate data on the system's losses, power under typical test conditions, and insolation in the system's plane must be obtained. In grid-connected photovoltaic systems, all system losses are combined for simplicity in the performance ratio, which is the ratio of the standard conditions module efficiency to the time-averaged system efficiency. The state-of-the-art performance ratio for grid-connected photovoltaic systems, which is currently typically 0.75–0.85, may eventually rise to 0.9. The typical performance ratio for modern standalone systems is 0.6.

4.2.8.2. Environmental Aspects

4.2.8.2.1. Environmental Life-Cycle Analysis

Although solar technologies don't produce emissions when they're in use, they can produce emissions during production and possibly during decommissioning (if they're made entirely by "solar breeders"). A lot of interest is currently being shown in the environmental aspects of solar technologies due to the industry's growth. Different materials and manufacturing processes for photovoltaic modules are already being developed as a result of environmental life-cycle analyses of solar systems and their components. One example would be creating pastes for screen printing that are based on water rather than organic solvents. Furthermore, a number of recycling procedures have been designed for modules that are rejected or off-spec.

4.2.8.2.2. Energy Payback Time

Whether or not the energy needed to manufacture a complete system is greater or less than the energy produced over its lifetime is one of the most contentious questions surrounding photovoltaics. In the beginning, photovoltaic systems were more energy consumers than producers. Stated differently, these systems had an energy payback period that exceeded their lifespan. Due to changes in circumstances, contemporary grid-connected rooftop photovoltaic systems now have payback periods significantly shorter than their (predicted) 30-year technical lifespan.

The materials used in the foundation and support cause the energy payback period of the system to be longer for grid-connected ground-based systems than for rooftop systems. The current three to nine-year energy payback period will drop to one or two years. The long energy payback period of the (lead-acid) battery makes the situation less favorable for stand-alone photovoltaic systems with battery storage (such as solar home systems) than for grid-connected systems. Modern solar home systems have an energy payback time of seven to 10 years at an annual insolation of 2,000 kWh/sq m. This figure might drop to about six years, five of which would be battery-related. The direct efficacy of (current generation) solar home systems for the reduction of greenhouse gas emissions is up for debate, as a photovoltaic system's battery typically has a technical lifespan of five years or less.

4.2.8.2.3. Carbon Dioxide Mitigation Potential

Based on energy payback time data, one can approximate the potential of photovoltaics to mitigate carbon dioxide emissions, provided that greenhouse gas emissions (SF_6 and CF_4) associated with the production of photovoltaic cells and modules are effectively minimized. For instance, a photovoltaic system with a technical lifetime of 30 years (ratio 1:15) and an energy payback time of two years at 1,500 kWh per square meter annually will generate 15 kWh of emission-free electricity for every kWh of electricity "invested" in manufacturing. As a result, specific carbon dioxide emissions are 15 times lower than those of the fuel mix that is utilized to supply energy to the entire photovoltaics industry chain.

4.2.8.2.4. Materials Availability

Thus far, the crystalline silicon photovoltaics sector has relied on off-grade semiconductor industry material as its feedstock. The crystalline silicon photovoltaics industry is growing at a very rapid rate, which means that "solar grade" silicon production needs to be dedicated. While several laboratory-scale processes for producing solar-grade silicon have been developed, none have been implemented into commercial production. However, it is anticipated that fresh feedstock will be accessible when needed. There are questions about the availability of some of the elements used in thin-film photovoltaic modules, such as tellurium and indium. There don't seem to be any short-term supply constraints, but at very large production levels (many gigawatts annually), the balance between the photovoltaic industry's demand and the global market's supply could become problematic. For this reason, CdTe, and CIGS could be useful bridging technologies (California Department of Water Resources, 2005; Towler, 2014).

4.3. ENERGY CONSERVATION

The goal of energy conservation is to use fewer energy services to minimize unnecessary energy consumption. Changing one's behavior to use less service, such as driving less, or using energy more wisely, which involves using less energy for continuous service, are two ways to achieve this. Efficient use of energy can lead to energy conservation and offers several benefits, such as decreased greenhouse gas emissions and carbon footprint, along with cost, water, and energy savings. Green engineering techniques extend the life of machine parts that transform energy into different forms.

Reducing waste and losses, increasing productivity through technology advancements, enhancing operations and maintenance, altering user behavior through user activities or profiling, keeping an eye on appliances, transferring load to off-peak hours, and offering energy-saving advice are all ways to conserve energy. One way to identify user habits and behaviors related to energy consumption is to monitor appliance usage, create an energy usage profile, and identify patterns of energy consumption when energy is used inefficiently. Energy-efficient appliances with high energy loads and consumption can be found with the aid of appliance energy profiling. Seasonal variations also have a significant impact on energy load because people use more heating in the winter and air conditioning in the summer. It is difficult but crucial to strike a balance between user comfort and energy load to preserve energy. Large-scale influences on energy consumption trends include economic growth, political unrest, technological advancements, and environmental concerns.

The physics principle of energy conservation states that in a closed system, the energy of interacting bodies or particles stays constant. Kinetic energy, or the energy of motion, was the first type of energy to be identified. When two particles collide, a phenomenon known as elastic collision occurs when the total kinetic energy of the particles before and after the collision is equal. The concept of energy has gradually expanded to encompass a wider range of forms. It was once believed that the kinetic energy lost by a body slowing down as it moves upward against gravity is transformed into potential energy, or stored energy, which is then transformed back into kinetic energy when the body accelerates on its way back to Earth. For instance, kinetic energy is transformed into potential energy when a pendulum swings upward. All of the system's energy is in potential energy when the pendulum momentarily stops at the peak of its swing, leaving zero kinetic energy. Potential energy is transformed back into kinetic energy when the pendulum swings back down. The total of kinetic and potential energy is constant at all times. Friction, on the other hand, causes even the most well-designed mechanisms to slow down and eventually lose energy. It was demonstrated beyond a reasonable doubt in the 1840s that the heat produced by friction could be included in the concept of energy. The total of thermal, potential, and kinetic energy is the quantity that is actually conserved. For instance, potential energy is transformed into kinetic energy when a block slides down a slope. The kinetic energy of the block is transformed into thermal energy when friction causes it to slow down. Energy simply transforms from potential to kinetic to thermal energy; it is neither created nor destroyed. The first law of thermodynamics is this application of the conservation of energy principle in its broadest sense. The idea of energy grew to encompass the energy of an electric current, the energy contained in fuels and other chemicals, and the energy stored in an electric or magnetic field. An automobile, for instance, moves when the kinetic energy of motion is transformed from the chemical energy in its gasoline.

It was not until the development of relativity physics in 1905 that mass was understood to be equal to energy. A system of high-speed particles' total energy consists of both their rest mass and the large mass gain they experience as a result of their high speed. The energy-conservation principle has also been referred to as the conservation of mass-energy or the conservation of total energy since the discovery of relativity.

Rather than rejecting the conservation principle, physicists accepted the existence of a new subatomic particle, the neutrino, that was supposed to carry off the missing energy when the principle seemed to fail, as it did when applied to the type of radioactivity known as beta decay (spontaneous electron ejection from atomic nuclei).

Later, an experimental detection of the neutrino was made. But energy conservation goes beyond being a universal principle that never goes out of style. The uniformity of time provides a mathematical demonstration of its follow-through. Energy would not be conserved if the same physical phenomena occurred at different times and required different amounts of energy if one moment of time were remarkably different from all other moments (California Department of Water Resources, 2005; Towler, 2014).

4.3.1. Importance of Energy Conservation

Energy conservation is critical for a number of reasons. It not only lessens our dependency on non-renewable energy sources (such as fossil fuels), but it also lowers our energy expenses, such as our utility and other energy bills. Energy conservation also reduces the amount of development that can be expanded in areas where the extraction of natural resources—like oil or even lithium—is affecting the environment. Greenhouse gas emissions into the Earth's atmosphere are directly decreased when we conserve energy and use it more wisely.

We are as dependent on energy as we are on air and water; we cannot imagine a moment in our lives without it. We utilize a variety of gadgets to ensure a comfortable life throughout the day. Naturally, there's nothing wrong with that. The moment we start wasting energy, the issue arises.

As was already mentioned, some people don't give much thought to the difficulties their future selves might face if they don't start acting with caution. They fail to consider how dire things would get if there were no more energy resources on Earth. The majority of the resources we use are non-renewable, and the renewable ones are not yet developed and widely available to be used. According to research, the energy we have left on Earth will only last for the next 40 years. It's time for us to start admitting that we have harmed the planet enough and to start acting appropriately to improve things. We can gradually begin to improve the current situation with reasonable energy use and appropriate conservation. To lessen the burden we place on the environment, energy conservation is now essential. Additionally, since our electricity bill will be significantly lower, energy conservation will also help us save money.

4.3.2. Measures for Energy Conservation

We can significantly impact the world by implementing minor adjustments in our daily routines. Every individual can contribute by avoiding the needless use of electronic devices like air conditioners, fans, lights, and other appliances.

The primary goal of National Energy Conservation Day is to increase awareness of people's lifestyle choices. These days, coal, natural gas, and other fossil fuels that form over millions of years are our main sources of energy. The fact that they are not renewable is the worst part. As a result, we need to concentrate more on producing energy using renewable resources, like biomass, geothermal, hydro, solar, and wind energy (California Department of Water Resources, 2005; Towler, 2014).

4.3.3. User Oriented Energy Conservation

Energy conservation is significantly impacted by user behavior. It includes the detection of user activity, profiling, and interactions between appliances. By identifying the user's energy consumption patterns, user profiling involves swapping out necessary system settings with automated ones that can be activated upon request. Personal traits play a key role in influencing energy-saving behavior in user profiling.

These attributes consist of social norms, age, gender, education, and household income. The influence of personality traits, social norms, and attitudes on energy conservation behavior is another factor that influences user behavior. Energy conservation behavior is influenced by attitudes and beliefs about convenient living, eco-friendly transportation, energy security, and residential location choices. Thus, implementing eco-friendly practices and energy-saving technologies can enable energy conservation. Having knowledge about energy-saving techniques can lead to prudent energy usage. The decisions made by users result in patterns of energy consumption. A thorough examination of these consumption patterns reveals waste energy patterns, and addressing those patterns could result in a notable reduction in energy load. Therefore, assessing the effects of energy-saving measures and finding solutions to environmental issues depend heavily on human behavior. Users' habit loops can be changed to achieve significant energy savings.

4.3.3.1. User Habits

Recommendations for bettering user habits contribute to energy conservation because they have a significant impact on energy demand. Realizing patterns in energy consumption requires understanding micro-moments, which are detected utilizing a range of sensing devices placed in conspicuous locations throughout the house. The micro-moment is an event that turns an appliance from inactive to active and aids in creating profiles of how much energy users use based on their activities. Energy conservation can be accomplished through user behavior, such as adhering to micro-moment energy-saving recommendations. The amount of energy wasted can be reduced by choosing an appropriate timetable for the use of appliances. It's important to comprehend user behavior with appliances to develop an efficient scheduling system.

4.3.3.2. Off-Peak Scheduling

One of the many energy-saving strategies is off-peak scheduling, which entails running an appliance during an energy-efficient hour. Once user habits regarding appliance use are understood, this schedule can be met. Setting an appliance to operate during an off-peak hour will greatly lower electricity bills because most energy providers split their energy tariff into high and low-price hours.

4.3.3.3. User Activity Detection

The accurate identification of the appliances needed for a task is made possible by the detection of user activity. An appliance can be turned off to save energy if it is not needed for the user's current activity and is running. To implement this energy-saving technique, user activities must be precisely identified.

4.3.3.4. World Energy Conservation

World Energy Conservation Day disseminates information about energy conservation and raises awareness of climate change and global warming. The main contributor to climate change is human activity. The disruption of the natural carbon cycle caused by the extraction and burning of fossil fuels (coal, gas, and oil) produces a massive amount of carbon dioxide, which warms the planet and is the root cause of the climate crisis. But the issue isn't with carbon. In addition to increasing pollution, we also clear forests, turn them into farmland, and construct cement- and concrete-based cities, which prevents the earth from naturally absorbing carbon dioxide. Reducing the amount of carbon dioxide emitters we use is one way to conserve energy resources (California Department of Water Resources, 2005; Towler, 2014).

4.4. ENVIRONMENTAL AND CARBON FOOTPRINT IMPLICATIONS

Hydropower is often touted as a renewable energy source, but its environmental and carbon footprint implications are complex and multi-faceted. Here's a detailed look at these implications:

4.4.1. Environmental Implications

1. Ecosystem Disruption
   - i. **Habitat Alteration:** Dams and reservoirs can significantly alter natural habitats, affecting local flora and fauna. Flooding caused by reservoirs can submerge forests, wetlands, and agricultural lands.
   - ii. **Aquatic Ecosystems:** The creation of a reservoir changes the natural flow of rivers, which can disrupt aquatic ecosystems. Species adapted to riverine conditions might struggle to survive in the altered environment.
   - iii. **Sedimentation:** Over time, sediment accumulation in reservoirs can impact water quality and reduce the lifespan of the dam. Sediment can also affect aquatic habitats downstream.
2. Fish Migration
   - i. **Barrier to Migration:** Dams can obstruct migratory routes for fish and other aquatic organisms, affecting their life cycles and population dynamics. Fish ladders and bypass systems can mitigate this but are not always fully effective.
   - ii. **Changes in Water Temperature and Chemistry:** Reservoirs can change water temperature and chemical composition, which might be detrimental to species adapted to the natural river conditions.
3. Greenhouse Gas Emissions
   - i. **Methane Emissions:** Reservoirs, especially in tropical regions, can produce significant amounts of methane due to the decomposition of organic matter in submerged soils and vegetation. Methane is a potent greenhouse gas.
   - ii. **Lifecycle Emissions:** While hydropower plants have low operational emissions, the construction, maintenance, and decommissioning phases can produce greenhouse gases. Emissions from concrete production (for dams) and steel (for turbines) are notable.

4. Impact on Local Communities

   i. **Displacement:** Large reservoirs can displace communities and affect their livelihoods. This displacement can lead to social and economic challenges for the affected populations.

   ii. **Changes in Water Availability:** Altered river flow can impact local water availability for drinking, agriculture, and other uses.

4.4.2. Carbon Footprint

1. Construction and Infrastructure

   i. **High Initial Carbon Footprint:** Building dams, reservoirs, and associated infrastructure requires significant amounts of concrete, steel, and other materials, leading to substantial carbon emissions.

   ii. **Operational Footprint:** Though hydropower has low direct operational carbon emissions, the lifecycle emissions from construction and maintenance are important to consider.

2. Maintenance and Operation

   i. **Energy Use:** Operating and maintaining hydropower facilities involve energy use, which can have associated carbon emissions depending on the source of the energy.

   ii. **Sediment Management:** Dredging and other sediment management activities can also have environmental and carbon costs.

3. Comparative Carbon Emissions

i. **Renewable Comparison:** Despite the environmental impacts, hydropower generally has a lower carbon footprint compared to fossil fuel-based energy sources. It is often used as a low-carbon complement to intermittent renewable sources like wind and solar.

4.4.3. Mitigation Strategies

1. Environmental Management

   i. **Fish-Friendly Technologies:** Implementation of fish ladders, bypass systems, and other technologies to minimize the impact on aquatic life.

   ii. **Environmental Flow Requirements:** Designing hydropower systems to maintain some level of natural flow in the river to support ecosystems downstream.

2. Greenhouse Gas Reduction
   i. **Improved Reservoir Management:** Techniques to reduce methane emissions, such as managing the decomposition of organic matter and promoting aeration.
   ii. **Energy-Efficient Technologies:** Using advanced, energy-efficient technologies for construction and operation to reduce overall carbon footprint.
3. Community Engagement
   i. **Stakeholder Involvement:** Involving local communities in planning and decision-making processes to address displacement and environmental concerns.
   ii. **Benefit Sharing:** Ensuring that communities benefit from hydropower projects through job creation, infrastructure improvements, and other forms of support.

A complicated range of environmental and carbon footprint issues are associated with hydropower. Despite being a lower carbon energy source than fossil fuels, it still has a big influence on the environment. Hydropower has many advantages, but there are also possible ecological and social costs. To balance these benefits, careful planning, technological innovation, and mitigation strategies are necessary (California Department of Water Resources, 2005; Towler, 2014).

4.4.4. Case Study: The Grand Coulee Dam and its Effectiveness in Reducing Emissions

4.4.4.1. Overview

The Grand Coulee Dam, located on the Columbia River in Washington State, USA, is one of the largest and most significant hydropower projects in the world. With a total installed capacity of 6,809 MW, the dam has been operational since 1942 and plays a critical role in providing renewable energy to the region.

4.4.4.2. Effectiveness in Reducing Emissions

1. **Greenhouse Gas Reduction:** The Grand Coulee Dam generates substantial quantities of low-carbon electricity, which significantly reduces reliance on fossil fuels such as coal and natural gas. By producing approximately 20 billion kilowatt-hours (kWh) of electricity annually, the dam helps offset the emissions that would otherwise be produced by fossil-fuel power plants. This shift

contributes to a notable decrease in greenhouse gas emissions. Estimates suggest that the dam's operation helps prevent around 10 million tons of CO_2 emissions annually, compared to the emissions from an equivalent amount of coal-fired power generation.

2. **Regional Impact:** The energy produced by the Grand Coulee Dam supports a large region, including major urban centers and industries. The availability of clean, renewable energy from the dam has facilitated a reduction in regional dependence on fossil fuels, helping to lower the overall carbon footprint of the Pacific Northwest. The integration of hydropower into the regional grid enhances energy security and stabilizes energy prices while promoting the transition to a more sustainable energy mix.
3. **Challenges and Considerations:** Despite its significant contributions to reducing emissions, the Grand Coulee Dam has associated environmental impacts. The dam's construction altered river ecosystems and impacted local fish populations, particularly salmon and other migratory species. Efforts have been made to mitigate these impacts through fish ladders and habitat restoration projects. However, balancing the environmental trade-offs with the benefits of clean energy remains a critical consideration.
4. **Long-Term Sustainability:** The Grand Coulee Dam's role in reducing emissions aligns with broader goals for sustainable energy development. Its ability to provide a consistent and reliable source of renewable energy supports ongoing efforts to address climate change and reduce reliance on fossil fuels. Additionally, the dam's capacity for future upgrades and integration with emerging technologies continues to enhance its effectiveness in contributing to a low-carbon energy future.

4.5. CONCLUSION

The Grand Coulee Dam exemplifies the effectiveness of large-scale hydropower projects in reducing greenhouse gas emissions. By displacing fossil fuel generation and providing a substantial amount of renewable energy, the dam plays a crucial role in the regional and national efforts to mitigate climate change. While the project presents environmental and ecological challenges, its contributions to reducing carbon emissions highlight its importance as a key component of a sustainable energy strategy (California Department of Water Resources, 2005; Towler, 2014).

REFERENCES

1. AdovorTsikudo, K. (2021). Ghana's Bui hydropower dam and linkage creation challenges. *Forum for Development Studies, 48*(1), 153–174. https://doi.org/10.1080/08039410.2020.1867714.
2. California Department of Water Resources. (2005). *File: OrovilleDam.jpg–Wikimedia Commons*. Oroville Dam. Retrieved from: https://commons.wikimedia.org/wiki/File:OrovilleDam.jpg (accessed on 9 November 2024).
3. Charlier, R. H. (2007). Forty candles for the Rance River TPP tides provide renewable and sustainable power generation. *Renewable and Sustainable Energy Reviews, 11*(9), 2032–2057. https://doi.org/10.1016/j.rser.2006.06.001.
4. Cheng, C., Chen, F., & Li, G. (2018). Reform and renewables in China: The architecture of Yunnan's hydropower-dominated electricity market. *Renewable and Sustainable Energy Reviews, 94*, 682–693. https://doi.org/10.1016/j.rser.2018.06.012.
5. Ehrbar, D., Schmocker, L., & Vetsch, D. F. (2018). Hydropower potential in the periglacial environment of Switzerland under climate change. *Sustainability, 10*, 2794. https://doi.org/10.3390/su10092794.
6. Fare, R., Grosskopf, S., & Pasurka, C. A. (2007). Environmental production functions and environmental directional distance functions. *Energy, 32*, 1055–1066. https://doi.org/10.1016/j.energy.2006.12.006.
7. Fernandes, G., Lima Gomes, L., & Teixeira Brandão, L. E. (2019). Mitigating hydrological risk with energy derivatives. *Energy Economics, 81*, 528–535. https://doi.org/10.1016/j.eneco.2019.05.022.
8. Firth, A., Zhang, B., & Yang, A. (2019). Quantification of global waste heat and its environmental effects. *Applied Energy, 235*, 1314–1334. https://doi.org/10.1016/j.apenergy.2018.10.008.
9. Førsund, F. R. (2014). Pumped-storage hydroelectricity. In *Hydropower Economics* (pp. 183–206). Springer.
10. Gottschalk, K. (2016). Hydro-politics and hydro-power: The century-long saga of the Inga project. *Canadian Journal of African Studies, 50*(2), 279–294. https://doi.org/10.1080/00083968.2016.1175920.
11. Harlan, T. (2018). Rural utility to low-carbon industry: Small hydropower and the industrialization of renewable energy in China. *Geoforum, 95*, 59–69. https://doi.org/10.1016/j.geoforum.2018.06.002.
12. Herrera-Estrada, J. E., Diffenbaugh, N. S., & Wagner, F. (2018). Response of electricity sector air pollution emissions to drought conditions in the

western United States. *Environmental Research Letters, 13*, 124032. https://doi.org/10.1088/1748-9326/aae501.

13. Kim, K. T., Lee, D. J., & Park, S. J. (2015). Measuring the efficiency of the investment for renewable energy in Korea using data envelopment analysis. *Renewable and Sustainable Energy Reviews, 47*, 694–702. https://doi.org/10.1016/j.rser.2015.03.016.
14. Koch, H., Chaves Silva, A. L., & Goncalves de Azevedo, J. R. (2018). Integrated hydro-and wind power generation: A game changer towards environmental flow in the Sub-middle and Lower São Francisco River Basin? *Regional Environmental Change, 18*, 1927–1942. https://doi.org/10.1007/s10113-018-1334-0.
15. Shokr, A. (2009). Hydropolitics, economy, and the Aswan High Dam in mid-century Egypt. *The Arab Studies Journal, 17*(1), 9–31.
16. Towler, B. F. (2014). Chapter 10–Hydroelectricity. In *The Future of Energy* (pp. 215–235). Academic Press.

Chapter 5

Biomass and Bioenergy

Contents

INTRODUCTION

Utilizing organic materials to produce electricity and lessen reliance on fossil fuels, biomass, and bioenergy constitute an essential component of renewable energy solutions. Bioenergy can be produced through a variety of processes using biomass, which is derived from plant and animal materials like wood chips, animal dung, and crop residues. One can use this energy to produce heat, electricity, or biofuels such as ethanol and biodiesel. Generally, one of three processes—gasification, anaerobic digestion, or combustion—is used to convert biomass into bioenergy. Each process produces a different type of energy and efficiency level. The potential for biomass to be carbon neutral is one of its main benefits. Carbon dioxide is taken up by plants during growth, and it is released back into the atmosphere when biomass is burned or breaks down. When compared to fossil fuels, this minimizes the net increase in atmospheric CO_2 and establishes a closed carbon cycle. Furthermore, by recycling organic matter, biomass can benefit waste management and soil health improvement by utilizing agricultural and forestry residues.

However, using biomass for environmental purposes requires sustainable practices. The benefits may be undermined by overharvesting, changes in land use, and ineffective conversion technologies, which can result in deforestation, habitat loss, and higher emissions. It is crucial to use ethical sourcing practices, optimize conversion technologies, and make sure biomass energy systems are carefully integrated with other renewable energy sources to optimize the benefits. All things considered, biomass, and bioenergy present a viable route to lower greenhouse gas emissions, improve energy security, and advance sustainable farming methods. Through technological advancements and the application of sustainable management practices, biomass has the potential to be a major contributor to a resilient and diversified energy future.

5.1. OVERVIEW OF BIOMASS AND BIOENERGY

Recent organic matter produced either directly or indirectly by living things is known as biomass. The process of photosynthesis, which uses energy from sunlight to transform carbon dioxide and water from the environment into the carbohydrates (sugars, starches, cellulose, and lignin) that make up the plant, is what produces the biomass in plant matter. Animals that consume plants and/or other animals, as well as the organic waste these animals produce, are examples of biomass. Thus, the different types of biomass store solar energy as chemical energy. The term "biomass" refers to a variety of materials that can be used to generate energy for human use, such as wood from plantation forests, waste from agriculture and the logging industry, vegetable, and animal fats, organic wastes from industry, animal husbandry, and human settlements, and residues from forest production and the logging industry. Biofuels can be either liquid or gaseous and are produced from a variety of organic materials known as biomass. While some biofuels—like wood and fibrous materials—can be used straight away as fuel sources with little processing, others must be indirectly obtained from biomass using the right conversion technology to provide energy in a form that is more convenient for providing the necessary energy service. The chemical energy contained in biofuels is commonly referred to as "bioenergy" (Wilkie, 2001; Graunke and Wilkie, 2008; Abatzoglou, 2009).

5.1.1. What Is Biomass?

Plant or animal matter used as fuel to generate heat or electricity is called biomass. Wood, energy crops, and waste from farms, yards, and forests are a few examples. Since biomass is technically capable of being used straight as fuel (e.g., wood logs), some individuals conflate the terms biofuel and biomass. Most of the time, biomass just refers to the biological raw material used to make the fuel. Typically, liquid or gaseous fuels used in transportation are referred to as biofuels.

The fuel developed from natural and organic materials or wastes, which are a renewable and sustainable source of energy is known as Biomass.

A few types of fuels used to generate biomass are:

1. Scrap lumber.
2. Woody construction and Forest debris (Wood waste, sawmill waste, green waste from landfills and other vegetative).
3. Certain agricultural crops and wastes.
4. Manure.
5. Animal waste.
6. Ethanol waste.
7. Municipal solid waste (sewage sludge or other landfill organics).
8. Landfill gas.
9. Other industrial waste (i.e., paper sludge from paper recycling processes).

It's a common misconception that the only energy sources available for producing biomass are plant wastes, but this is untrue. Materials derived from animals are also utilized to generate biomass. The primary component of biofuels, biomass is widely utilized in the production of electricity and heat in the energy context as a substitute for fossil fuels. The elements that makeup biomass are alkaline earth, heavy metals, nitrogen, hydrogen, and oxygen in addition to alkali and nitrogen atoms.

5.1.1.1. Difference Between Biomass and Other Fossil Fuels

1. The first, foremost, and major difference is the "Time Difference," that is, the time required for it to be generated.
2. Biomass takes carbon out of the atmosphere while it is developing, and returns as it is burned.
3. Biomass can be converted into ethanol by a thermal process and into methanol by fermentation and digestion.
4. Fuels have a high energy density, and burning and utilizing it releases carbon dioxide into the atmosphere (Figure 5.1).

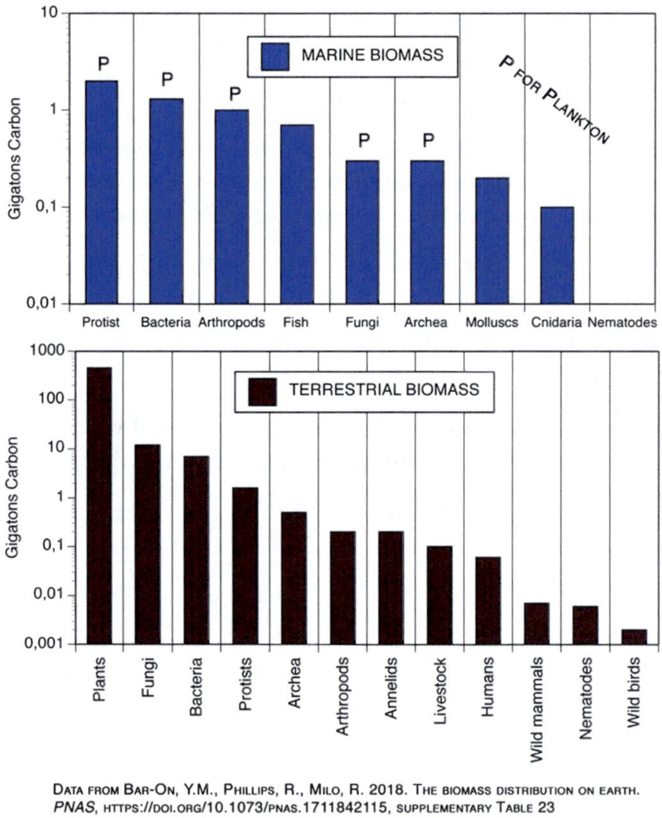

Figure 5.1. *Different types of biomass.*

Source: By John Richard Dolan – The Journal of Plankton research, using data from supplementary matter Bar-On, Yinon M., Rob Phillips, and Ron Milo. "The biomass distribution on Earth." Proceedings of the National Academy of Sciences (2018): 201711842., Attribution, https://commons.wikimedia.org/w/index.php?curid=69671555.

5.1.1.2. Disadvantages of Biomass

Agricultural wastes wouldn't be discovered if crops weren't grown. It is necessary to develop harvesting techniques if there is a demand for biomass. The fact that land used to grow crops for biomass is permanently occupied and cannot be used for other uses, such as housing, conservation, farming, resorts, or agricultural use, is another significant drawback.

5.1.2. Biomass System

1. For the industrial sector, Sigma Thermal offers fully packaged biomass systems. Our systems are designed, engineered, and manufactured with decades of energy experience in mind. Electric automation controls, an ash handling system, and a fuel feed system are all included in packaged biomass energy systems. For the duration of the system, components, training, installation supervision, and equipment service management are offered to support and maintain effective operation (Figure 5.2).

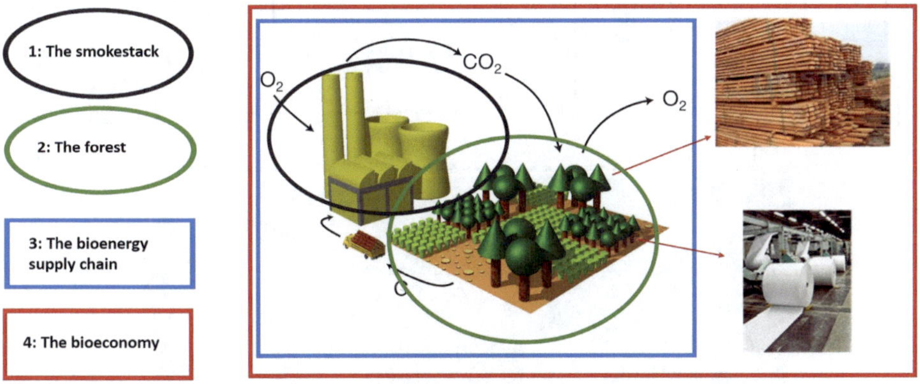

Figure 5.2. *Biomass heating system.*

Source: By Annette L. Cowie, Göran Berndes, Niclas Scott Bentsen, Miguel Brandão, Francesco Cherubini, Gustaf Egnell, Brendan George, Leif Gustavsson, Marc Hanewinkel, Zoe M. Harris, Filip Johnsson, Martin Junginger, Keith L. Kline, Kati Koponen, Jaap Koppejan, Florian Kraxner, Patrick Lamers, Stefan Majer, Eric Marland, Gert-Jan Nabuurs, Luc Pelkmans, Roger Sathre, Marcus Schaub, Charles Tattersall Smith Jr., Sampo Soimakallio, Floor Van Der Hilst, Jeremy Woods, Fabiano A. Ximenes – https://onlinelibrary.wiley.com/doi/10.1111/gcbb.12844, CC BY 4.0, https://commons.wikimedia.org/w/index.php?curid=113261191.

5.1.2.1. Reception Tank

Two reception tanks are usually used: one for other types of biomass, like organic industrial waste, and one for slurry and manure. The reception tank serves as a buffer tank to guarantee that the plant continues to operate on the weekends and on holidays. To guarantee that the biomass supplied to the reactor is uniform, various biomasses are additionally blended in the receiving tanks. Slurry reception tanks are concrete tanks that are covered and can hold enough material to last for seven days. Industrial waste reception tanks frequently have bigger capacities.

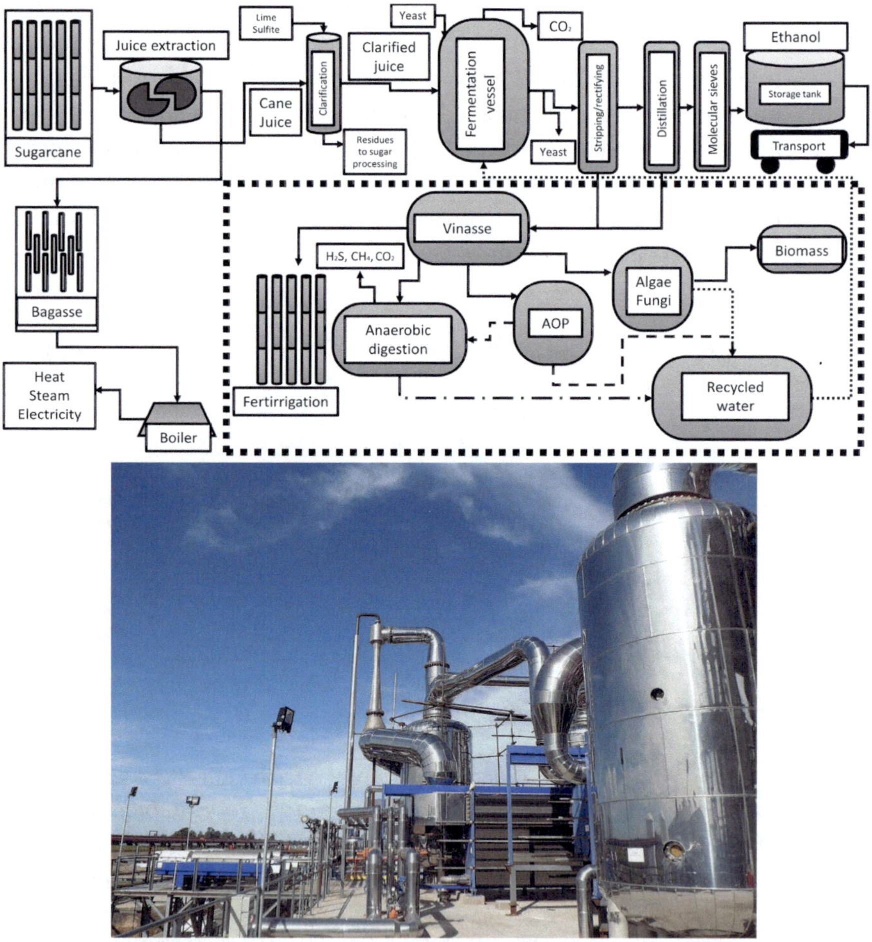

Figure 5.3. *Digestion of "Vinasse"/residual material from ethanol production.*

Source: Rodrigues Reis CE and Hu B (2017) Vinasse from Sugarcane Ethanol Production: Better Treatment or Better Utilization? Front. Energy Res. 5:7. doi: 10.3389/fenrg.2017.00007.

The tanks are thoroughly mixed to avoid the accumulation of layers, and occasionally the industrial waste receiving tank is heated to guarantee, for example, that the fat portion remains liquid. Large amounts of sand and grit may settle out in the slurry reception tanks and need to be periodically removed. A mixer tank may be positioned in between the reactor and the receiving tanks in certain systems. The reception tanks may release a variety of smelly substances. These can be removed and cleaned using various air purification techniques (Figure 5.3).

5.1.2.2. Biomass Feeding Pump

A pump moves the biomass from the reception tank to the reactor tank. The pump can be immersed in the reception tank or sit in its own pump well. The pump often has an associated comminutor, which shreds the biomass.

5.1.2.3. Reactor Tank

The reactor tank is either an airtight-sealed concrete tank or a fully enclosed, insulated steel tank. The biomass that is being broken down can be heated by heating coils installed in the tank, or an external heat exchange system can be used as the heat source. A stirrer built into the tank can fully agitate the entire volume, preventing the formation of a surface crust. It also has pressure and temperature gauges, an overflow outlet, and other features. Lastly, a high-pressure valve makes sure that if the gas removal fails, the pressure doesn't rise too high. An outlet for the generated biogas is located at the top of the tank. Generally speaking, the reactor tank holds 10–20 times the daily biomass input for a thermophilic process and 15–25 times the daily biomass input for a mesophilic process.

5.1.2.4. Effluent Discharge Pump

A pump moves the digested biomass to a storage tank. In simple plants, and with a suitable arrangement of pipes and valves, one pump can take care of both the feeding and discharge.

5.1.2.5. Digestate Storage Tank

The storage tank, also known as the secondary digester, serves as a holding tank until the digested biomass is either transported to the farmer's storage tank or applied directly to the farm as fertilizer. Usually, the tank is covered to keep ammonia from escaping and to stop rainwater from entering. Since the biomass in a fully stirred reactor can never be fully digested, a certain amount of gas will also be produced from the storage tank if the retention period is extended. It is also possible to extract and use this gas (Wilkie, 2001; Graunke and Wilkie, 2008; Abatzoglou, 2009).

5.2. BIOMASS RESOURCES

Native vegetation, forest plantations and forestry residues, wood, and wood processing wastes, agricultural crops and crop residues from field and processing operations, herbaceous, and woody energy crops, wastes resulting from various animal husbandry activities, food processing wastes, industrial organic wastes, vegetable oil, and municipal solid wastes are some of Australia's abundant biomass resources. Except wood wastes that result from processing native forest products and are allowed to be used as a biomass resource, native forests and old growth forests are not taken into account as biomass resources for energy production. These forests' features are also not further discussed in this handbook (Figure 5.4).

The biomass resources that are considered eligible biomass resources under the renewable energy:

- Forest and forestry residues.
- Wood and wood wastes.
- Agricultural crop residues.
- Agricultural process residues
- Energy crops.
- Black liquor.
- Wet waste from animal husbandry and food processing.
- Municipal solid wastes.

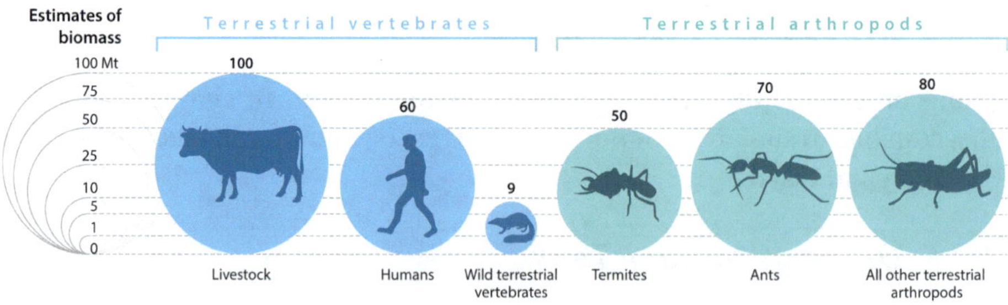

Figure 5.4. *Biomass includes plants, animals, and microorganisms.*

Source: By Paul Eggleton – "The State of the World's Insects": Figure 2., CC BY 4.0, https://commons.wikimedia.org/w/index.php?curid=101266909.

5.2.1. Biomass Utilization Technologies

There are numerous technological avenues via which biomass resources can be transformed into beneficial energy service products. The abundance of potential and current biomass resources in Australia has led to the commercial development and use of a variety of biomass utilization technologies.

Biomass utilization technologies can be classified based on either (i) the fuel conversion process or (ii) the energy service product. The primary energy conversion processes are:

1. Combustion in which the biomass is burnt in the presence of excess air to release its energy as heat. Furnace technology options for combustion include underfed grates, traveling grates, pulverized fuel burners, cyclone burners, and fluidized beds.
2. Gasification in which the biomass is burnt in a reduced-air atmosphere so that it is converted into a product gas rich in carbon monoxide and hydrogen. The gasification process can be performed in fluidized beds and fixed beds.
3. Pyrolysis which is the thermal degradation of biomass fuel in the absence of oxygen. The biomass is converted into a range of products comprising char, fuel gas and liquid "bio-oil." The bio-oil seems to be the attractive value-added product in this process and the recent trend is the development of "fast pyrolysis" technology which has a higher yield of bio-oil that can be used as a chemical feedstock or as a fuel oil in a furnace. The fluidized bed is the preferred technology for the pyrolysis reaction.
4. Biochemical processes which can also be used to convert biomass into methane or liquid fuels such as alcohols by anaerobic digestion and fermentation processes. Because of the complex molecular structure of some biomass materials, hydrolysis is required before fermentation in those cases (Figure 5.5).

Based on the energy service product, biomass utilization technologies can be classified into those for producing:

- Heat and electricity.
- Biogas.
- Biofuels.

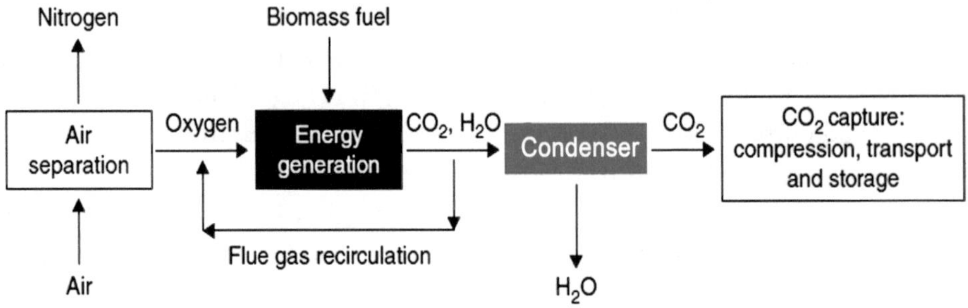

Figure 5.5. *Processes for the conversion of biomass to fuels, energy, and heat.*

Source: By Clair Gough-University of Manchester, UKPatricia Thornley-University of Manchester, UKSarah Mander-University of Manchester, UKNaomi Vaughan-University of East Anglia, Norwich, UKAmanda Lea-Langton-University of Manchester, UK – Biomass Energy with Carbon Capture and Storage Unlocking Negative Emissions, CC BY-SA 4.0, https://commons.wikimedia.org/w/index.php?curid=74791655.

5.2.2. Biomass Production Potential and Efficiencies

Biological energy production is ultimately limited by photosynthetic efficiencies. Nowadays, the efficiency of forestry and agriculture rarely surpasses 1% of the total solar energy input transformed into a product that can be recovered. Nonetheless, photosynthesis has the enormous potential to produce food, fiber, fuel, and forest products even at current productivity levels. Overall primary productivity is orders of magnitude higher than current fossil fuel consumption. A recent group of experts predicted that better agricultural practices, forest preservation and management, and soil carbon sequestration could reduce the greenhouse effect by storing carbon dioxide in soil and long-term biomass and by substituting biofuels for fossil fuels. They came to the conclusion that biological systems could offset all of the CO_2 emissions caused by fossil fuels. Furthermore, this could be done without reducing the amount of food, fiber, or forest products produced. According to this group's even more conservative estimates, biological processes could produce greenhouse gas offsets at comparatively low costs that would be nearly half the current annual increase in atmospheric CO_2. The heroic planetary management measures advocated by some, like increasing primary productivity through trace nutrient fertilization of terrestrial and oceanic environments, are not even necessary for these actions. Naturally, these broad estimates need to be balanced with more thorough evaluations of the particular local and regional biomass energy system possibilities. For instance, it is assumed in the majority of biomass system projections for greenhouse gas mitigation that photosynthesis-fixed carbon can be stored for a very long time (> 100 years) in soil carbon, above-ground biomass, and other long-term storage forms. This is problematic because, as forests reach their maximum (climax) biomass levels, it will become a limiting factor in a few decades. Therefore, the creation of biological energy systems that use biomass fuels in place of fossil fuels to reduce emissions of CO_2 and other greenhouse gases must be the primary focus. Maximizing the effectiveness of the first phase of the processes + photosynthesis is essential to such systems.

Even the most fruitful outdoor experiments have not been able to demonstrate more than half of this in practice, and in plant cultivation, the maximum conversion efficiency is only 1 to 3%, of which less than half is usually recovered in the final product (grain, fiber, etc.). Due in large part to the minimal inputs (fertilizers, pest control, and management) into forestry, much lower productivity levels for trees have been observed.

Results from experimental and commercial systems for microalgae systems, which have extremely high productivity, do not differ significantly from those of higher plants grown in similar climates but with higher productivity. However, a 10% solar conversion efficiency ought to be achievable if laboratory results using microalgae cultures could be attained in outdoor ponds, that is, roughly a three-fold increase in productivity over the current level, or more than 200 metric tons/hectare/year.

The reason why such high solar productivity and efficiencies have not been attained yet is that at high light intensities, for example sunlight, the quantity of chlorophyll molecules that absorb a photon far outweighs the ability of the photosynthetic machinery to utilize this abundance of trapped energy. Algae near the surface waste light while those below are shaded, leading to a decrease in productivity. The excess photons are lost as heat or fluorescence. Even though this issue was identified and potential solutions were put forth decades ago, there hasn't been much advancement in this field for several years.

Prisms or optical fibers that carry light to the ponds' lower layers are one suggested remedy. While these devices are workable in lab environments, their use is prohibitively expensive. Another strategy is to quickly mix the culture so that, on average, the algae are moved to the dark where they can process the photon energy more slowly after it has been absorbed by the photosynthetic apparatus.

But this also needs mixing so quickly that it's merely a curiosity for laboratories (though of great fundamental interest). The third strategy is to lessen the algae's chlorophyll content to prevent or at least minimize photon absorption and consequently light waste and self-shading. Modern genetic engineering techniques have the potential to achieve a lower chlorophyll content. This method offers the clearest way to achieve high photosynthetic efficiencies, almost to the theoretical maximum, with algal cultures grown outdoors in full sunlight.

However, it must be acknowledged that this would only be the first—albeit a very significant—step in the larger endeavor to create highly productive, reasonably priced microalgal production systems. The creation of algal strains with high yields of desired products—such as starches for ethanol fermentation, vegetable oils, hydrocarbons, or hydrogen—is a further step. Once more, we can use the capabilities of contemporary molecular biology to achieve these objectives. In particular, the quickly developing field of metabolic engineering, which permits the manipulation and control of entire enzyme pathways, is one such technology. The simultaneous optimization of photosynthetic efficiencies and product formation may be possible through metabolic engineering.

Similar to higher plant agriculture, where grain formation occurs after vegetative biomass production, a two-stage process would probably be ideal. Just as crop plants cannot compete with weed species, low chlorophyll algal strains and, more generally, metabolically engineered strains that yield high levels of desired products, could not compete with wild-type algae. Algal ponds are susceptible to disease and grazer incursions, just like agricultural systems. To stop, or at least reduce, such contamination and enable open-pond mass culture of specific species—especially genetically/metabolically engineered strains—new techniques for microecological engineering must be developed. Closed photobioreactors, the alternative method, are only feasible a small-scale inoculum production stage. As explained below, the development of the other biological energy conversion processes covered in this report, including the fermentations of ethanol and methane, will also require microecological engineering—the fusion of microbial ecology with fermentation and molecular biotechnologies.

The unexpected discovery that mutants of the green alga Chlamydomonas still performed both reactions at nearly normal rates despite lacking half of the photosynthetic machinery typically involved in CO_2 fixation and hydrogen production. This implies that algal cultures might actually need fewer photons than the 8–10 quanta required by accepted photosynthesis mechanisms to reduce one CO_2 molecule to carbohydrates. While proof of such elevated photosynthetic efficiencies is still pending, this serves as a reminder that discoveries continue to emerge even at the most basic levels of our comprehension of biological processes. This further illustrates how challenging it is to forecast any future technology. In any case, it seems feasible to aim for significantly higher than present efficiencies with both microalgae and higher plants, outside in the sun, which supports sustained and targeted R&D. The knowledge and methods created to maximize the productivity of microalgae mass cultures may also be used to increase plant photosynthesis. While light saturation limits are typically not as severe as those found in algal ponds, there is still a great deal of room to improve the productivity of higher plant production systems. The use of such techniques in forestry would be of greatest potential, both in terms of yield improvements and increasing global biomass resources.

5.2.3. Fuel Alcohol Production from Biomass

The synthesis of ethanol from biomass may be one of the most significant and immediate uses of biological energy systems. Within the U.S. currently, fermentations using corn starch as the substrate produce more than 5 billion liters of ethanol fuel annually. This is a conventional technology that is only financially viable due to significant government subsidies that promote the production of corn and its subsequent conversion to fuel ethanol. With the growing need for food in the world, some process improvements are possible, but no significant discoveries are anticipated, nor is the price of corn starch expected to decrease. The energy and greenhouse gas emissions balances for com-to-ethanol systems are also not very good; at most, they are only marginally positive, and even then, only with the use of more costly, capital-intensive technology. It is necessary to use less expensive fermentation processes and substrates (Figure 5.6).

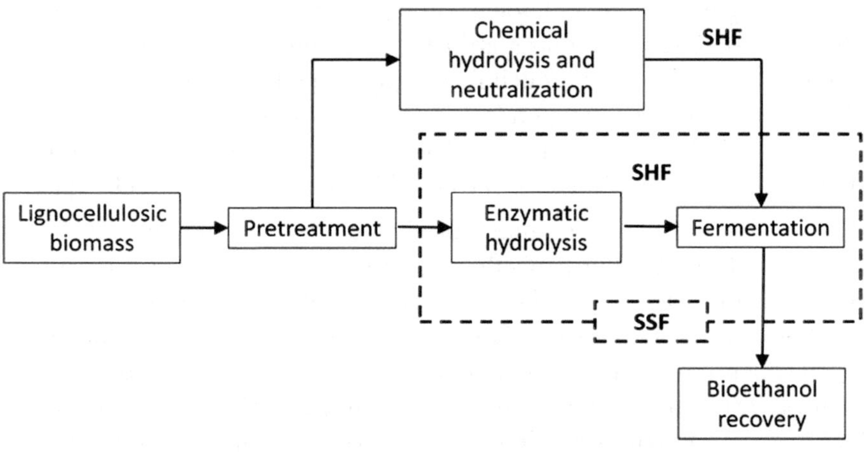

Figure 5.6. *Basic pathway for ethanol production from lignocellulosic biomass.*

Source: Tuong An Tran, T., Kim Phung Le, T., Phong Mai, T., and Quan Nguyen, D. (2020). Bioethanol Production from Lignocellulosic Biomass. IntechOpen. doi: 10.5772/intechopen.86437.

First, the biomass must be pretreated to release the lignocellulosic polymers and facilitate their hydrolysis by acids or enzymes. This is followed by the fermentation of the pentose and hexose sugars, which are obtained from the hemicellulose and cellulose fractions, respectively. The remaining lignin can be used as a boiler fuel or as a feedstock for products with higher value, such as polymers and phenolics). Decades of research have been spent in this area, but no workable procedures have yet been shown. In fact, several issues impede these processes: the laborious pretreatment; the slow and large-enzyme-requiring enzyme hydrolysis; the rapid and high-degradation product-producing acid hydrolysis that yields low yields; and the two types of sugars present that complicate the fermentation processes themselves, as well as the possibility of contamination. Regarding contamination, most demonstration projects and even laboratory research use sterilizable fermenters that are impractical for large-scale commercial operations. Contamination in traditional corn starch fermentations is reduced by the batch fermentation process's relative speed, specific conditions (e.g., low pH), vaccination, and hygienic (but not sterile) operating and startup conditions. The main issue with lignocellulosic ethanol fermentation is the lengthy enzymatic scarification process, which is chosen over acid hydrolysis because it produces higher yields. One of the main limitations of the enzyme hydrolysis process is the exponential increase in contamination issues with cycle time.

The focus of recent research has been on using genetically modified microbes to develop better procedures for the enzymatic hydrolysis and fermentation of pentose sugars. Microecological engineering techniques will need to be developed to maintain such strains in large systems, subject to invasion and contamination by potentially much faster-growing wild microbes, as in the case of microalgae culture in open ponds. By using such microecological techniques, the need to strictly maintain (expensive) aseptic conditions or maximize the amounts and activities of the enzymes used in this process would be relieved.

Future lignocellulosic biomass production is expected to cost as little as \$0.13/liter (\$0.50/gallon) of ethanol, as opposed to nearly three times as much for existing processes. These low-cost estimates are predicated on many favorable hypotheses, one of which is that the cellulolytic enzymes would be provided concurrently by the same microbial culture, which would also ferment the entire suite of carbohydrates produced by enzymatic hydrolysis. This gets rid of the expensive step of producing separate fermentation stages for the process. Even though these projections are extremely optimistic, they highlight the crucial research and development goals needed to develop the extremely low-cost processes that would make this technology competitive with fossil fuels as well as with less expensive biomass conversion methods like gasification and combustion.

In the near future, the most easily fermentable and least expensive resources—such as waste from paper mills and food processing facilities—will probably be used to develop the technology for making ethanol from cellulosic biomass. Agricultural residues have a high ash content, which makes them less suitable for thermochemical conversion processes. However, over time, they could be a significant source of ethanol. But this still necessitates a significant investment in R&D. Projects for planned demonstration in the U.S. should make it possible to determine how far technology has come and how far it might still need to go. The relative market shares of the different lignocellulosic biomass energy conversion technologies—combustion, gasification, pyrolysis, ethanol fermentation, and, subsequently, methane fermentation—are hard to forecast. Regardless of the technology used for conversion, lignocellulosic biomass will undoubtedly remain a significant renewable energy source in the future—if not the primary one—for the vast majority of people on the planet (Wilkie, 2001; Graunke and Wilkie, 2008; Abatzoglou, 2009).

5.2.4. Methane Fermentations

Nowadays, methane fermentations—which have been reviewed—are frequently employed to convert municipal sludge (e.g. wastes from food processing) and certain industrial to biogas, a carbon dioxide and methane blend. Some developed countries (Denmark, for example) also turn some animal manures into biogas, and some less developed nations (China and India, in particular) use inexpensive farm-scale systems that are fed agricultural and animal wastes. Furthermore, during the last 10 years, many cutting-edge procedures for the use of low-strength wastes have been commercialized. Even with these uses, only a tiny portion of the resources that could be used for methane fermentations are being used.

One explanation for this is that methane fermentation (also known as anaerobic digestion) processes have advanced more quickly than biogas utilization technologies. Taking into account the costs of gas clean-up (to remove CO_2 and other impurities) and compression, replacing natural gas with biogas is typically not practical due to its relatively small production. Generally speaking, it is not practical to store biogas for longer than a day or so. Biogas must therefore be used primarily as and where it is produced. The best option for medium-sized to large facilities is to generate electricity using biogas. But to do this, generators are needed, which add approximately $0.025/kWhr to the cost of electricity after accounting for capital and operating expenses. This is on top of the cost of the biogas. Alternatively, biogas can be compressed moderately and most of the CO_2 removed before being used as a fuel for vehicles. Numerous American towns have employed these systems the U.S, and can offer reasonably priced transportation fuels for specific uses (e.g., fleets of local services). These strategies, however, are only appropriate in situations where there are ample, continuous waste resources and high local expenses associated with using conventional power or transportation fuels.

Smaller-scale systems must combine the use of biogas with residential or light commercial/industrial applications, such as water heating and cooking. In addition to using biogas, the technology for producing it still needs to be improved, especially for wastes and residues that are high in solids and lignocellulosic content. Similar to ethanol, agricultural residues offer a sizable potential resource for the production of biogas. High solid fermentation processes still need to be developed, though. Generally speaking, anaerobic digestion of municipal solid waste using the highly mechanical systems currently in use in Europe is not feasible, unless cost is (nearly) nonexistent. One option to consider is utilizing batch pits enclosed with plastic liners for low-cost, high-solids methane fermentations. Their usefulness has yet to be proven, though. There is still work to be done on controlling gas production to better match biogas production with utilization, as well as adding moisture, nutrients, and adapted microbial cultures to prevent volatile acid overproduction (which would inhibit the process) and rate limitations in such systems.

When compared to the other technologies this report highlights, biogas production is undoubtedly the most straightforward and possibly least expensive method of producing biological energy. Any mitigation (capture, use) of anthropogenic sources of methane, such as from animal wastes, is even more valuable than other biological energy sources, which reduce CO_2 emissions by replacing fossil fuels. This is because methane gas has a large greenhouse effect, about 10 times that of CO_2. Based on this advantage, biogas technologies—such as the recovery of gas from animal manure and landfills—should be given more thought.

The development of micro-ecological engineering techniques for methane fermentations is necessary to achieve maximal gas production in a predictable and controlled manner, better matching production to utilization. This management concept should also include hydrogen fermentations, either for the production of pure hydrogen or for the production of methane-hydrogen mixtures for vehicle fuels (a volumetric mixture of roughly 70:30 CH_4:H_2 significantly reduces tail pipe emissions). The creation of process control methods for covered lagoons or high solids batch processes—systems that offer methane fermentations the greatest potential in the future but also the most technological difficulties—will be the bigger task.

5.2.5. Fuels Derived from Microalgae

Since the late 1950s, microalgae mass cultures have been explored as possible sources of biofuel. The first conceptual engineering analysis was presented in this regard. Microalgae systems are relatively undeveloped and require longer-term research and development (R&D) compared to other biological energy systems, either combined with waste conversion or as stand-alone systems (growing of energy crops followed by utilization as, or conversion to, fuels) (Figure 5.7).

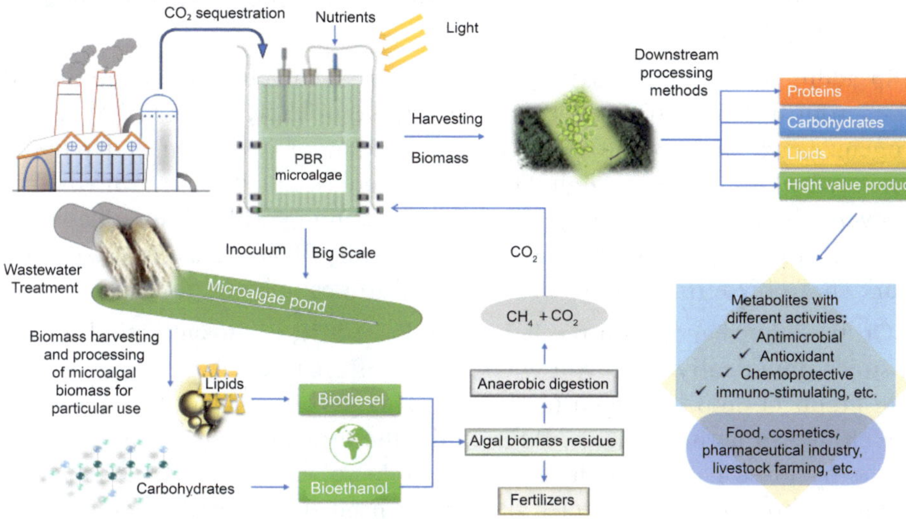

Figure 5.7. *Potentials of microalgae.*

Source: Ilieva Y, Zaharieva MM, Kroumov AD, Najdenski H. Antimicrobial and Ecological Potential of Chlorellaceae and Scenedesmaceae with a Focus on Wastewater Treatment and Industry. Fermentation. 2024; 10(7):341. https://doi.org/10.3390/fermentation10070341.

When it comes to the production of hydrogen by microalgae, there are two main approaches that can be taken into consideration: direct processes, which involve transferring electrons from water to hydrogen directly, and indirect processes, which start with fixing CO_2 in a fermentable substrate and subsequently use algae metabolism to convert it to hydrogen. Among other drawbacks, the former method is limited by the hydrogen evolution reaction's sensitivity to oxygen, whereas the latter is more intricate, necessitating several steps and demanding metabolic regulation. This concept has economic potential, according to a recent economic analysis of a large-scale (over 100 hectares) two-stage indirect process, with costs estimated at $12 to 15 per million BTU (million British thermal units). Direct bio photolysis systems would be more expensive, as the entire surface area would need to be covered in closed photo bioreactors, vs. merely 10% of this for a two-step procedure. The aforementioned cost estimate was predicated on ambitious but reasonable long-term R&D goals: a 10% solar conversion efficiency in the first stage (CO_2 fixation) and a 100% solar conversion efficiency in the second stage (H_2 evolution). The majority of the energy in the second stage is derived from starch that has been stored, which enables such high solar efficiencies. Naturally, while based on established metabolic processes in microalgae, this process is still primarily conceptual at this point.

In addition to photosynthesis, the processes involved in the evolution of hydrogen must also be effective. This practically completely removes the enzyme nitrogenize, which is required for the production of hydrogen and reduces efficiency by half due to its high ATP requirement. It is still unknown whether the reversible hydrogenate enzyme can produce high yields, despite having a much higher intrinsic efficiency. Yields are typically 20% or less stoichiometric in dark anaerobic fermentations of organic substrates to hydrogen.

On the other hand, with the help of light energy, photosynthetic bacteria can stoichiometrically convert a wide variety of organic substrates, including waste products, into hydrogen (and CO_2). This process has attracted a lot of research due to its high yields. Regretfully, the utilization of the nitrogenase enzyme by these bacteria reduces the possible solar conversion efficiencies. Similarly, nitrogen-fixing species have been used in the majority of studies on cyanobacteria's indirect hydrogen production. Microalgae that generate hydrogen through a reversible hydrogenate ought to be given preference in the future. Regardless, it will take time and effort to produce hydrogen through photosynthetic processes. The creation of closed, inexpensive photobioreactors that would house the light-driven hydrogen evolution reactions is one of the many difficulties.

Similarly, one long-term goal of R&D is to produce lipids, or vegetable oils and hydrocarbons, using microalgae. It is possible to envision a two-stage process that is essentially the same as the one thought up for hydrogen production: the first stage would involve the production of algal biomass, and the second stage would involve lipid induction. Nonetheless, compared to the closed photo bioreactors envisioned for the stage of hydrogen production previously discussed, both stages would be open ponds and significantly less expensive. Lipids would be converted to biodiesel (methyl or ethyl esters) or used directly as fuel for vehicles. The CO_2 needed to grow the algae would come from flue-gases from power plants or other comparable sources. As a result, using this procedure would be a direct way to lower greenhouse gas emissions from power plants. In comparison to other options like ocean disposal, microalgae ponds offer one of the more affordable ways to directly utilize and mitigate CO_2 emissions from power plants, even though they are probably more expensive than the majority of other biomass energy systems.

Over the past few decades, microalgae biotechnology has come a long way from being a lab curiosity to relatively large-scale (> 10 hectares) commercial systems based on open ponds that produce several thousand tons of biomass annually. But we're still in the early stages of this technology. Due to chemical selection, only two species of algae—Spirulinaplatensis and Dunaliellasalina—can be grown outdoors with confidence when given the right conditions (high bicarbonate and salinity, respectively). Although successful at the 0.2-hectare demonstration scale, the cultivation of high-lipid microalgae in open ponds still needs to be developed as a workable procedure. Nearer-term development and applications of microalgae processes for renewable energy production are likely when these processes are combined with other goals, specifically waste water treatment and the production of higher-value products, such as foods and chemicals.

All of our food, forest products, and even a sizable portion of our fuel and fiber needs are met by plant and algal photosynthesis, which powers the global biological carbon cycle, which far outpaces the carbon cycle from fossil fuels. Biomass systems have a lot of potential for significant growth in terms of renewable resources and green energy. The main source of this expansion needs to be higher productivity, that is, solar conversion efficiencies, but without adding more chemicals, fertilizers, or other non-renewable inputs to the farming systems, as has happened throughout the agricultural revolution.

Because they illustrate the basic methods for boosting photosynthetic efficiencies, microalgae can direct the path toward achieving these goals. The development of micro ecological engineering methods—which are necessary to attain process stability and low costs in open microbial processes for the production of hydrogen, ethanol, and methane from conventional biomass sources—can also be facilitated by microalgae production systems. When compared to specialized, single-purpose energy farming systems, microalgae processes can show how multipurpose/multiproduct systems maximize environmental and economic benefits by combining waste treatment with the production of higher value products and renewable fuels. The economics of biomass energy products in comparison to other sources' competitive costs will determine their future viability. Because fossil fuels are currently so cheap, biomass energy sources are typically not economically competitive.

However, the low cost of extracting fossil fuels is not the only reason for their low prices; perhaps more importantly, the prices of fossil fuels do not accurately reflect the true costs to society, the environment, and our future. The external costs associated with fossil fuel resources, such as environmental degradation, climate change, and the unsustainability of our economic systems, fall on all of us, but especially on future generations. While biomass energy systems can already rival fossil fuels in certain situations, a significant increase in the use of renewable bioenergy will necessitate advancements in both technology and economic accounting, given that biomass energy systems have lower costs to society and the environment than fossil fuels.

Bioenergy systems do, of course, have their own real and potential environmental effects. The largest-scale human interventions in natural ecosystems, which frequently result in their complete destruction and loss of their benefits—from biodiversity to hydrological cycle management—are found in agriculture and managed forestry. Large-scale, intensive biomass systems may have environmental effects that are greater than even those of fossil fuels, adding insult to injury. But there is an alternative: the creation of integrated, multifunctional, sustainable biomass systems that are based on highly effective microbial and photosynthesis processes and produce a range of goods, including fuels. In addition to supplying the food, fiber, and even a large portion of the fuel needed for all of humanity to have a decent standard of living in the twenty-first century, such systems have the potential to lessen the negative environmental effects of forestry and agriculture (Wilkie, 2001; Graunke and Wilkie, 2008; Abatzoglou, 2009).

5.3. BIOGAS ENERGY

Biogas is a mixture of gases that can catch fire. It is created when organic compounds, such as methane (CH_4) and carbon dioxide (CO_2), are broken down anaerobically by bacteria, devoid of oxygen. The gases that are produced are byproducts of these decomposer microorganisms' respiration, and the makeup of the gases varies depending on the material being broken down. CH_4 production is minimal if the material is mostly composed of high-molecular compounds (polymers) like cellulose and hemicellulose and carbohydrates like glucose and other simple sugars. On the other hand, methane production increases with a high-fat content (Figure 5.8).

Figure 5.8. *Pipes carrying biogas (foreground) and condensate.*

Source: https://upload.wikimedia.org/wikipedia/commons/f/f3/Biogas.jpg.

CH_4 – and whatever additional hydrogen there maybe – makes up the combustible part of biogas. $CH_4$is a colorless and odorless gas with a boiling point of –162□ and it burns with a blue flame. CH_4 is also the main constituent (77–90%) of natural gas. Chemically, CH_4 belongs to the alkanes and is the simplest possible form of these. At normal temperature and pressure, CH_4 has a density of approximately 0.75 kg/m^3. Due to CO_2being somewhat heavier, biogas has a slightly higher density of 1.15 kg/m^3. Pure CH_4 has an upper calorific value of 39.8 MJ/m^3, which corresponds to 11.06 kWh/ m^3. If biogas is mixed with 10–20% air, you get explosive air, which – as the name indicates – is explosive.

i. **Heating (calorific) value:** The quantity of heat produced at a complete combustion of a unit of a combustible compound.

ii. **Upper Heating Value:** The quantity of heat produced when the fumes are cooled to the extent that the water content (both that of the fuel and the quantity produced in the combustion process) condenses and releases its latent heat.

iii. **Lower Heating Value:** The quantity of heat produced when the water in the fuel does not vaporize and release its latent heat.

5.3.1. What Is Biogas?

One kind of biofuel that is naturally produced when organic waste breaks down is called biogas. CH_4 and CO_2 are the main gases released when organic matter, such as food scraps and animal dung, decomposes in an anaerobic environment (one without oxygen). Anaerobic digestion is another name for the process of producing biogas because this breakdown takes place in an anaerobic environment (Figure 5.9).

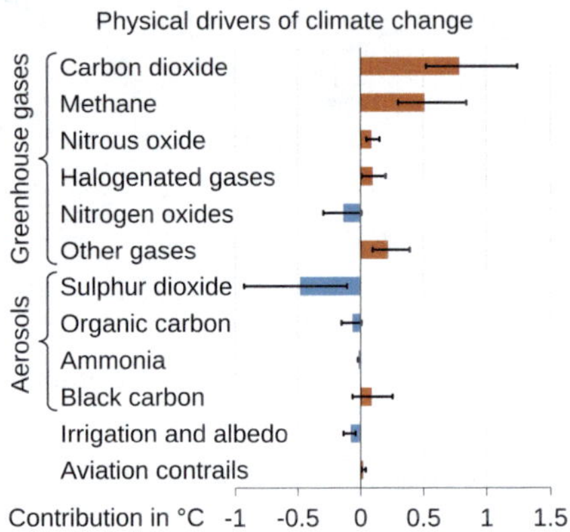

Figure 5.9. *Global warming and independence of fossil fuels, biogas has a number of advantages. Produced from biodegradable wastes, biogas can be used for the production of low-carbon heat and electricity and can be upgraded for use as a vehicle fuel.*

Source: By Eric Fisk – Own work, CC BY-SA 4.0, https://commons.wikimedia.org/w/index.php?curid=81034563.

Anaerobic digestion is a naturally occurring waste-to-energy method that breaks down organic matter through fermentation. Anaerobic digestion can produce biogas from organic matter such as animal manure, food scraps, wastewater, and sewage. Biogas can be used as an energy source because of its high CH_4 content, which is usually between 50 and 75%. Biogas is flammable and produces a deep blue flame as a result.

5.3.2. Ecology of Biogas

Biogas is known as an environmentally-friendly energy source because it alleviates two major environmental problems simultaneously:

- The global waste epidemic that releases dangerous levels of CH_4 gas every day.
- The reliance on fossil fuel energy to meet global energy demand.

Biogas utilizes nature's elegant tendency to recycle substances into productive resources by turning organic waste into energy. By treating waste on-site, biogas generation recovers waste materials that would otherwise contaminate landfills and avoids the use of hazardous chemicals in sewage treatment plants. It also saves money, energy, and materials. Furthermore, using biogas to generate energy does not require the extraction of fossil fuels. Rather, biogas transforms a hazardous gas into one that is far safer.

More precisely, CO_2 is produced from the CH_4 found in decomposing waste. The capacity of CH_4 gas to trap heat is roughly 20–30 times greater than that of CO_2. This indicates that a bread loaf's environmental impact will be roughly 10 times lower when it turns into biogas from rotting than if it were allowed to rot in a landfill (Figure 5.10).

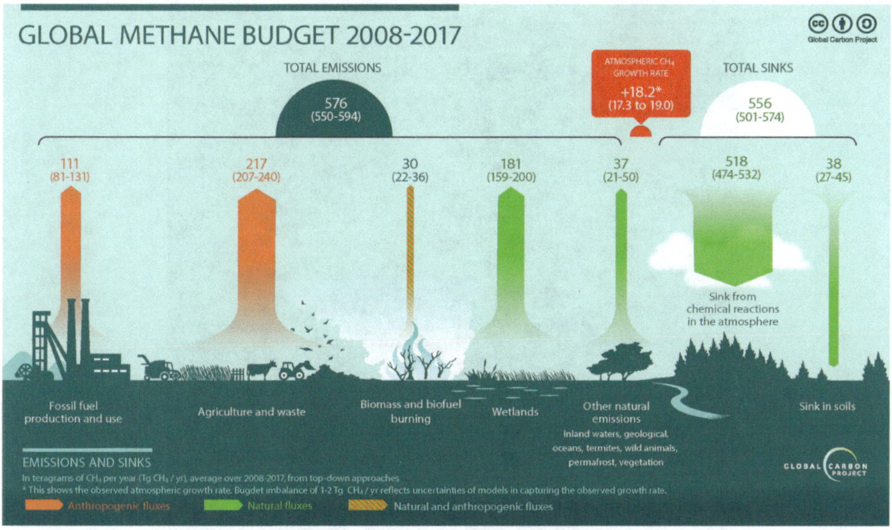

Figure 5.10. *The release of methane through human activity is intensifying climate change.*

Source: By The Global Carbon Project – http://www.globalcarbonatlas.org/en/CH4-emissions/https://essd.copernicus.org/articles/12/1561/2020/, CC BY-SA 4.0, https://commons.wikimedia.org/w/index.php?curid=93292720.

5.3.3. Biogas Digesters

Biogas digesters are the systems that convert waste into biogas and then direct that biogas so that the energy can be used productively, as opposed to allowing methane gas to escape into the atmosphere. Numerous plant types and biogas systems have been developed to utilize biogas effectively. Although the input, output, size, and type of each model vary, the biological process that turns organic waste into biogas is consistent. Organic matter is fed to biogas digesters, where it breaks down in a digestion chamber. Since the digestion chamber is completely submerged in water, it lacks oxygen and is an anaerobic environment. The organic material can be broken down and transformed into biogas by microorganisms in the anaerobic environment (Figure 5.11).

Figure 5.11. *A home biogas digester is essentially, a wet, enclosed compost pile that's setup to capture the methane released as organic waste decomposes.*

Source: https://commons.wikimedia.org/wiki/File:Samuchit_Household_Biogas_Plant_-_Urban.jpg.

5.3.3.1. All-Natural Fertilizer

The organic material breaks down in a liquid environment, allowing nutrients from the waste to dissolve into the water and form a nutrient-rich sludge that is usually applied to plants as fertilizer. Because it is produced daily, this fertilizer output is an extremely fruitful by-product of anaerobic digestion (Figure 5.12).

Figure 5.12. *These materials and other types of natural fertilizers for plants can be found at most garden centers.*

Source: By Alandmanson – Own work, CC BY-SA 4.0, https://commons.wikimedia.org/w/index.php?curid=49622141.

5.3.3.2. Biological Breakdown

To produce biogas, organic matter ferments with the help of bacterial communities. Four stages of fermentation move the organic material from its initial composition into its biogas state.

1. The first stage of the digestion process is the hydrolysis stage. In the hydrolysis stage insoluble organic polymers (such as carbohydrates) are broken down, making it accessible to the next stage of bacteria called acidogenic bacteria.

2. The acidogenic bacteria convert sugars and amino acids into carbon dioxide, hydrogen, ammonia, and organic acids.
3. In the third stage, the acetogenic bacteria convert the organic acids into acetic acid, hydrogen, ammonia, and carbon dioxide, allowing for the final stage- the methanogens.
4. The methanogens convert these final components into CH_4 and CO_2- which can then be used as a flammable, green energy.

5.3.4. Uses of Biogas

Different kinds of organic matter can be used to produce biogas, so different biogas digester models exist. Municipal wastewater, industrial wastewater, municipal solid waste, and agricultural waste are among the waste types that certain industrial systems are intended to handle. Animal waste is usually digested in small-scale systems. Furthermore, more recent family-size systems are made to break down food waste. The resulting biogas can be used as fuel for transportation, as well as gas, electricity, and heat. For instance, refined biogas powers hundreds of automobiles and buses in Sweden. In Sweden, landfills, and sewage treatment facilities are the main sources of biogas (Figure 5.13).

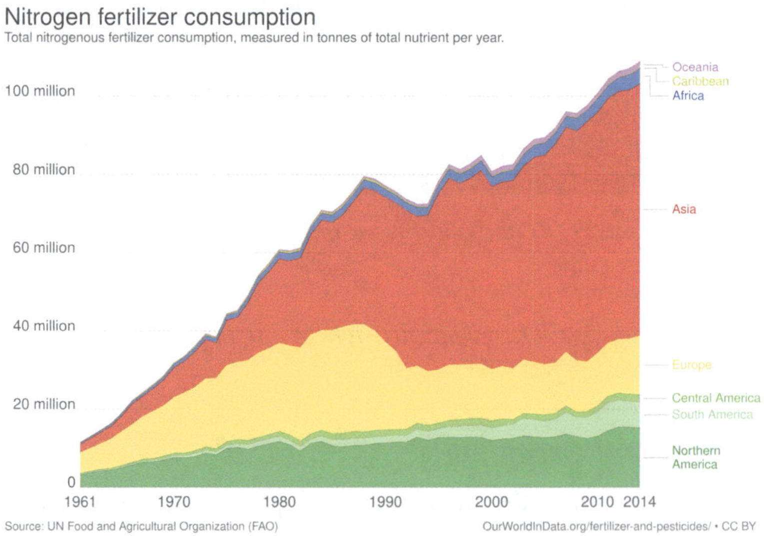

Figure 5.13. *Farm uses biogas-fueled gensets for cost-efficient power.*

Source: By Our World In Data – https://ourworldindata.org/grapher/nitrogen-fertilizer-consumption, CC BY 3.0, https://commons.wikimedia.org/w/index.php?curid=86924698.

5.3.4.1. Small-Scale Biogas Systems

China and India are the countries where small-scale, family-sized biogas digesters are most commonly found. However, because of more sophisticated and practical technologies like Home Biogas, the demand for these units is rising quickly throughout the world. People are eager to find environmentally friendly ways to handle their trash because the modern world is producing an increasing amount of waste.

Conventional systems, mainly in China and India, concentrate on animal waste. The combination of an excess of animal manure and a lack of energy in rural areas has made biogas digesters extremely popular, practical, and even life-changing. Because of the many advantages that using biogas produces, the government and local ministries of many developing nations even support and encourage the use of biogas digesters. Not only do biogas digesters provide clean, renewable energy for cooking, but many families also use the fertilizer by-product they produce extensively (Figure 5.14).

Figure 5.14. *Small-scale biogas production (using plastic bladders and tanks).*

Source: By F Moreau Lille3 – Own work, CC BY-SA 3.0, https://commons.wikimedia.org/w/index.php?curid=8973314.

Selling the bio-slurry byproduct that biogas systems produce is one way that some biogas users even make money. This bio-slurry is not the same as the daily produced liquid fertilizer. The term "bio-slurry" describes the state of organic matter that has undergone the most breakdown after being broken down by the system. When bio-slurry accumulates, it sinks to the bottom of the biogas system and can be easily emptied out with the aid of contemporary units like Home Biogas (usually an annual process). In actuality, this bio-slurry is a nutrient-rich sludge that enriches soil with many benefits and can boost vegetable garden productivity.

Technology imitating nature's capacity for replenishment is called biogas. Biogas units, whether industrial or family-sized, are growing in popularity and significance in the modern world. Biogas has the potential to significantly reduce greenhouse gas emissions as its application and efficiency increase. Biogas is useful in both developed and developing nations as a clean energy source and a sustainable way to handle organic waste.

5.3.5. Role of Biogas Digester

A biogas digester, sometimes referred to as a CH_4 digester, is a piece of machinery that can convert organic waste into fuel that is useful. Biogas digesters help to dispose of waste that would otherwise be discarded and offer low-cost fuel to those living in poverty in addition to being a source of renewable fuel. Numerous countries have made investments in the study of biogas digesters, which range from single-family-use devices to large-scale industrial equipment capable of producing significant amounts of electricity.

The biogas digester is dependent on the microbial breakdown of biomass, which is biological waste that includes everything from cow dung to kitchen scraps. Anybody who has strolled past an untended outhouse or compost pile knows that when anaerobic conditions arise in a biomass collection, they draw bacterial organisms that release various unique gases during the digestion process, chief among them being CH_4. These gases can be very helpful, but they are typically seen as an indication of inefficiency and are vented off for disposal (Figure 5.15).

Anaerobic fermentation and digestion are actually promoted in a biogas digester, and the gases are released into a storage container. Burning the resultant biogas can be used to generate electricity, heat, and cook. Fertilizer can be made from the sludge that is left over from the fermentation process in the biogas digester. Biogas digesters are sometimes referred to as anaerobic digesters because the process they use is anaerobic in nature. The removal of agricultural and human waste is one of the main uses for biogas digesters.

Figure 5.15. *Tank design plays an important role in optimizing the biogas output from anaerobic digestion.*

Source: By Volker Thies (Asdrubal) – Own work, CC BY-SA 3.0, https://commons.wikimedia.org/w/index.php?curid=2200301.

Energy availability and waste management are two major problems in many developing countries. Organizations and the government could help improve public health while simultaneously giving communities access to a sustainable energy source by supplying biogas digesters. Biogas digesters can be used by farms, which often produce a lot of animal waste, to power their operations or to produce electricity that can be sold or traded.

The biogas digester is one solution that proponents of sustainable energy have highlighted as a way to lessen dependency on fossil fuels. Biogas digesters not only produce sustainable energy but also have low costs, high efficiency, and little maintenance requirements. Because of this, they are a great resource for organizations that work to empower communities by giving them tools they can use and maintain themselves, as opposed to just providing aid, which can lead to communities becoming dependent on outside help.

5.3.6. Advantages and Disadvantages of Biogas

Interest in renewable energy sources is growing on a global scale. As more people set up biogas plants to produce biogas, the amount of biogas produced is increasing steadily. We put together this list of benefits and drawbacks of biogas so that you can see more clearly what it is good for (Figure 5.16).

Figure 5.16. *Wastewater treatment process produces sewage sludge, which is de-watered and dried either mechanically or thermally. In certain cases, the thickened sludge is digested.*

Source: By Cec-clp – Own work, CC0, https://commons.wikimedia.org/w/index.php?curid=14914191.

5.3.6.1. Advantages of Biogas

1. **Biogas is Eco-Friendly:** Clean and renewable energy can be obtained from biogas. Biodigestion produces non-polluting gas, which actually lowers greenhouse emissions (i.e., lessens the impact of the greenhouse). Because there is no combustion involved in the process, no greenhouse gasses are released into the atmosphere, making the use of gas from waste as a source of energy an excellent way to fight global warming.

Naturally, one of the main causes of the increased use of biogas is environmental concern. Because biogas plants capture and use methane, a harmful gas, as fuel, they significantly reduce greenhouse gas emissions. The production of biogas reduces the need for fossil fuels like coal and oil.

The fact that biogas is a naturally occurring process that doesn't require energy for generation is another benefit over other forms of renewable energy.

Additionally, since crops and trees will always grow, the raw materials used to produce biogas are renewable. It's a very sustainable option because the raw materials—manure, food scraps, and crop residue—are always available.

2. **Biogas Generation Reduces Soil and Water Pollution:** Not only do overflowing landfills release unpleasant odors, but they also let poisonous liquids seep into subterranean water sources. Therefore, the potential for biogas generation to enhance water quality is yet another benefit of biogas. Anaerobic digestion also effectively lowers the incidence of waterborne illnesses because it deactivates pathogens and parasites. Similarly, regions with biogas plants have much better waste collection and management. Improvements in the environment, sanitation, and hygiene follow from this (Figure 5.17).

Figure 5.17. *Municipal solid waste (MSW). Such land pollution can contaminate the soil and water and is a health hazard to local communities.*

Source: By Cezary p – Own work, CC BY-SA 4.0, https://commons.wikimedia.org/w/index.php?curid=4092940.

3. **Biogas Generation Produces Organic Fertilizer:** Digestate, a byproduct of the biogas generation process, is an excellent addition to or replacement for chemical fertilizers. While commercial fertilizers contain chemicals that have toxic effects and can cause food poisoning among other things, the fertilizer discharge from the digester can accelerate plant growth and resilience to diseases.

4. **It's A Simple and Low-Cost Technology That Encourages A Circular Economy:** The biogas production technology is relatively inexpensive. On a small scale, it requires little investment and is simple to set up. Animal dung and kitchen scraps can be utilized in tiny bio-digesters that you can use at home. After a while, a home system pays for itself, and the materials needed to generate it are completely free. The gas that has appeared can be used straight away to cook and create electricity. This explains the comparatively low cost of biogas production. Biogas plants and the waste products that their livestock produce on a daily basis can be utilized by farms. One cow's waste materials can generate enough energy to run a lightbulb for the whole day. Biogas can also be compressed in large plants to a natural gas-like quality and used as an automotive fuel. Such plants can be built for relatively little money, and doing so generates green jobs.
5. **Healthy Cooking Alternative for Developing Areas:** Women and children are spared the difficult task of gathering firewood thanks to biogas generators. More time is thus available for cleaning and cooking. What's more, cooking on a gas stove keeps the family safe from kitchen smoke exposure rather than over an open flame. By doing this, fatal respiratory illnesses are avoided. Sadly, the inefficient use of solid fuels for cooking results in home air pollution, which claims the lives of 4.3 million people annually.

5.3.6.2. Disadvantages of Biogas

1. **Few Technological Advancements:** The inefficiency of the systems used to produce biogas nowadays is a regrettable drawback of biogas. No new technologies have emerged to date to streamline the procedure and make it affordable and widely available. This indicates that it is still not feasible to produce on a large scale to feed a large population. Even though there are currently biogas plants that can meet some energy needs, many governments are unwilling to make investments in this industry (Figure 5.18).
2. **Contains Impurities:** Even after compression and refinement, contaminants are still present in biogas. The engine's metal components may corrode if the produced biofuel is used to power cars. Maintenance expenses would go up as a result of this corrosion. Lighters, water boilers, and kitchen stoves are far better suited for the gaseous mixture.

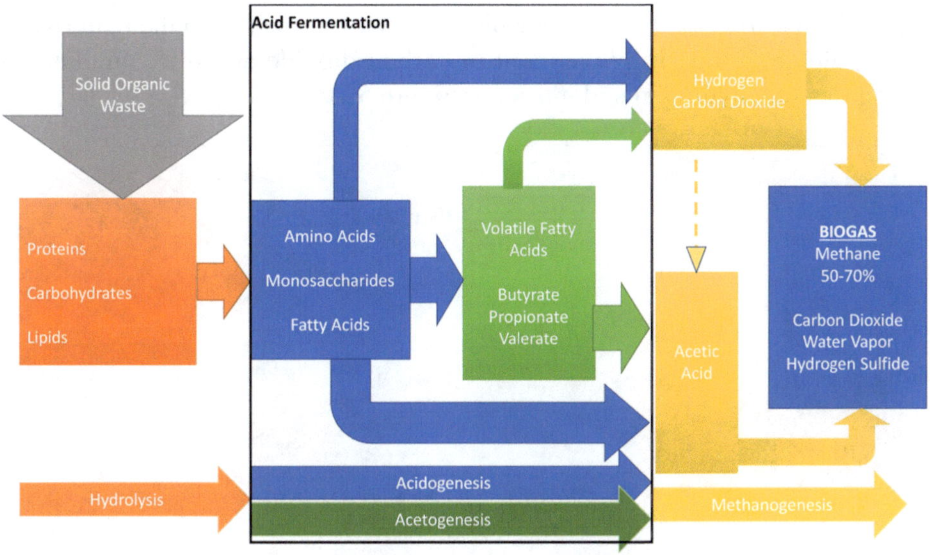

Figure 5.18. *Homebiogas is an off-grid system that generates clean energy without any electricity and allows you to properly treat your household waste. The system produces up to 2 hours of cooking gas every day solely from your food scraps or solid waste.*

Source: By Dwyerbm – Own work, CC BY-SA 4.0, https://commons.wikimedia.org/w/index.php?curid=117680440.

3. **Effect of Temperature on Biogas Production:** Much like other forms of renewable energy (e.g., the weather has an impact on biogas generation from solar and wind energy. About 37□ is the ideal temperature for bacteria to break down waste. Digesters need heat energy to keep the flow of biogas steady in cold climates.
4. **Less Suitable For Dense Metropolitan Areas:** Another biogas disadvantage is that industrial biogas plants only make sense where raw materials are in plentiful supply (food waste, manure). For this reason, biogas generation is much more suitable for rural and suburban areas (Wilkie, 2001; Graunke and Wilkie, 2008; Abatzoglou, 2009).

5.4. BIOGAS TECHNOLOGY

Biogas technology, as it relates to renewable energy, describes systems intended to convert organic waste products into energy that can be used. An organic substance like manure, plant matter, or even municipal waste materials can be anaerobically processed to produce biogas. Methane makes up the majority of biogas, along with a sizable amount of carbon dioxide and trace amounts of other gases like hydrogen and nitrogen (Figure 5.19).

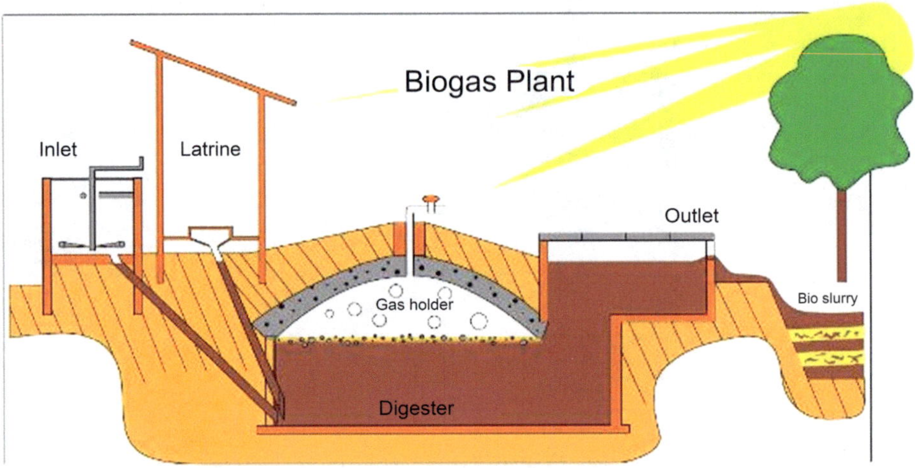

Figure 5.19. *The delivery of two kompogas digesters for a biogas plant in Chongqing marks the successful entry of Zurich-based Hitachi Zosen Inova into the Chinese market.*

Source: https://en.wikipedia.org/wiki/File:Biogas_plant_sketch.jpg.

Biogas energy can be used as an alternative to fossil fuels because it is a flammable material that burns similarly to liquefied petroleum gas (LPG). A biogas plant, a device that breaks down organic materials into gas, is frequently used to produce biogas. Farms frequently employ biogas technology with the main objective of reducing waste pollution. For instance, there may be serious issues with manure contaminating water sources on dairy farms. Farmers can solve this issue with the added benefit of producing a new power source thanks to biogas technology.

On certain farms, biogas technology systems enable the local production of biogas electricity and energy for heating. A farmer with 500 cows, for instance, could set up a system to process the manure the cows produced and produce enough electricity to run the entire dairy farm. In certain nations, more expansive systems gather manure from several farms and handle the digestion process centrally in a sizable facility (Figure 5.20).

Figure 5.20. *Liquefied petroleum gas (LPG) storage tanks solution.*

Source: By Manjarul.hasan – Own work, CC BY-SA 4.0, https://commons.wikimedia.org/w/index.php?curid=95495208.

Sewage treatment may make use of a similar biogas technology. Anaerobic digestion is one method for breaking down organic solids found in wastewater. Wastewater contains organic matter. The amount of dangerous bacteria in the water may also be lowered by this procedure, which is also known as sludge digestion. Digesters using biogas technology come in three primary varieties: plug-flow, complete mix, and covered lagoon. These kinds all break down manure to create biogas. Their effectiveness, the amount of solids they can manage, and their applicability to various environments and uses vary.

The most basic kind of digesters are covered lagoon ones, which just have a storage pool for the manure and a cover to trap the gas. This kind of technology works best with liquid manure that contains a small amount of solid matter, and it is appropriate for warm climates. Complete mix digesters use a mechanical mixing device to accelerate the digestion process while storing manure in a heated tank. More sophisticated biogas technology systems called plug-flow digesters actively move the manure being processed through the system while it is being broken down.

5.4.1. Biogas Process

The full biological breakdown of organic matter in the absence of oxygen, producing carbon dioxide (CO_2) and methane (CH_4) that is, anaerobic digestion is a complex process involving the cooperation of several distinct bacteria, each of which is in charge of completing a specific task. The bacteria are interdependent because what is considered a waste product by some may act as a substrate or food for other bacteria. The anaerobic process has a much lower energy yield than the aerobic (oxygen-rich) breakdown of organic matter. For instance, the aerobic breakdown of glucose produces a net yield of 38 ATP molecules, whereas the anaerobic breakdown of the same material produces only 2 ATP molecules. This indicates that the production of biomass (in the form of living bacteria) is lower per gram of decomposed organic matter and that the growth rate of anaerobic bacteria is significantly lower than that of aerobic bacteria. Under anaerobic conditions, the yield of 1 g of substance decomposes into 0.1 g of biomass, compared to 0.5g in aerobic conditions. Three steps are commonly identified in the biogas process: hydrolysis, acidogenesis, and methanogenesis. Each step is carried out by a distinct group of bacteria (Figure 5.21).

Figure 5.21. *To use biogas in any industrial process, it is important that the gas is filtered, or conditioned, prior to use.*

Source: By Unknown author – https://catalog.hathitrust.org/Record/003495329, Public Domain, https://commons.wikimedia.org/w/index.php?curid=104771654.

5.4.2. Hydrolysis

Long-chain molecules like fat, protein, and carbohydrate polymers hydrolyze to produce monomers or small molecules. The breakdown is catalyzed by several distinct enzymes produced by various specialized bacteria, and it occurs extracellularly. that is,, it occurs in the surrounding liquid outside the bacterial cell. Under anaerobic conditions, proteins, simple sugars, and starch hydrolyze readily. Other polymeric carbon compounds break down a little more slowly, but lignin—a crucial component of plants—never breaks down in anaerobic environments (Figure 5.22).

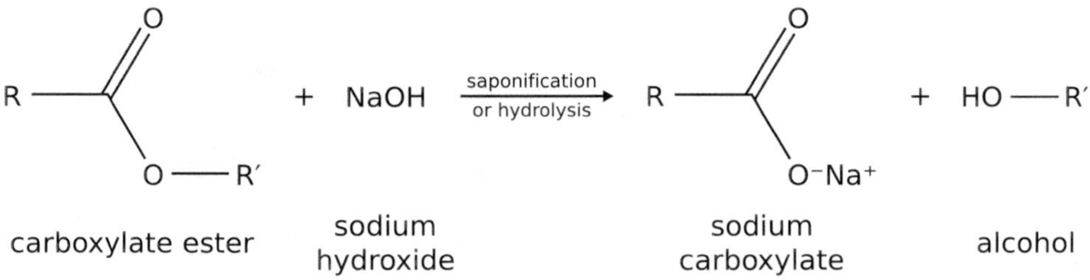

Figure 5.22. *Reaction of methyl acetate and water demonstrating the hydrolysis of an ester.*

Source: https://commons.wikimedia.org/wiki/File:Ester_hydrolysis.svg.

Complex polysaccharides that are readily hydrolyzed by certain bacteria are cellulose, a polymer made up of several glucose molecules, and hemicellulose, a polymer made up of several other sugars. Because cellulose and hemicellulose are firmly packed in lignin in plant tissue, bacteria have a hard time getting access to them. For this reason, only roughly. The biogas process breaks down 40% of the cellulose and hemicellulose in pig slurry. In general, the rate at which organic matter breaks down into CH_4 and CO_2 is not constant; for animal dung and other substrates with a high concentration of complex molecules, this rate is typically only between 30 and 60%.

5.4.3. Fermentation – Acidogenesis

About 50% of the monomers (glucose, xylose, amino acids) and long-chain fatty acids (LCFA) are converted to acetic acid (CH_3COOH) in a balanced bacterial process. 30% is broken down into short-chain volatile fatty acids (VFA), and the remaining 20% is converted to hydrogen (H_2) and carbon dioxide (CO_2).

Monocarboxylic acids, or fatty acids, are present in fats. The majority of fatty acids found in nature have an even number of carbon atoms. Less than six carbon atoms make up VFAs. More than six carbon atoms make up LCFAs (Figure 5.23).

Figure 5.23. *Anaerobic digestion is a biological process by which the organic matter from various origins is turned into biogas for energy production.*

Source: By Vortexrealm at English Wikipedia, CC BY 3.0, https://commons.wikimedia.org/w/index.php?curid=7929120.

If there is an imbalance, the slow-growing VFA-degrading bacteria will not be able to keep up, increasing the relative level of VFAs and the risk of accumulation and the process "turning sour." Thus, a consistent breakdown of VFAs is essential and frequently a process limiter for biogas production. Simple fats hydrolyze to produce 1 mol glycerol and 3 mol LCFA. Larger amounts of protein, which contain nitrogen in amino groups (-NH_2), will therefore produce large amounts of ammonium/ammonia (NH_4^+ /NH_3), whereas larger amounts of fat in the substrate will result in large amounts of LCFA. In both situations, this may inhibit the next stage of decomposition, especially if the biomass feedstock's composition changes.

5.4.4. Methanogenesis

Methanogens, also known as methanogenic bacteria, are responsible for producing CH_4. Methanogens are members of the kingdom Archaea, which is a taxonomic class that also includes bacteria and eukaryotes. Since a kingdom is the highest taxonomic level, archaea belong to the same taxonomic category as bacteria (Eubacteria), fungi, plants, and animals. Methanogens are thought to have been among the planet's first living things.

The two distinct bacterial groups are in charge of producing CH_4. While one group converts acetic acid to CH_4, another uses hydrogen and carbon dioxide to create CH_4. About 70% of the methane produced under stable conditions comes from the breakdown of CH_3COOH, with the remaining 30% coming from hydrogen and carbon dioxide. Because of the delicate balance between the two processes, inhibiting one will inevitably inhibit the other.

Among the bacteria involved in the process, methanogens grow at the slowest rate. As a result, the amount of material that can be digested and the rate at which the process can proceed are both limited by them. Methanogens grow at a rate that is only about 25% faster than that of acid-forming bacteria. In the process, not much energy is released by the methanogens. However, because of the anoxic conditions, there is less competition from other bacteria, which allows them to survive.

5.4.5. Process Parameters for a Biogas Plant

For a biogas process to be effective and productive, there are a number of parameters that have to be optimized (Table 5.1) (Table 5.2).

Table 5.1. *Biogas Top-Down Modeling (Energy Systems, Regional Impact, Greenhouse Gas (GHG) Emission): Please Refer to the Research Article for the Back References*

View on Biogas Plants	Coding/ Software	Biomass Modeling	Region	Additional Modeling
Energy system	oemof	Annual chemical biogas potential	Northwestern Germany	Time-dependent electricity production (wind and photovoltaic) and demand
Energy system	oemof	Annual chemical biogas potential	Northwestern Germany	Time-dependent electricity production (wind and photovoltaic) and demand
Energy system	Engineering Equation Solver	daily chemical biogas potential based on chicken manure and maize silage	-	Electrical energy production (wind, photovoltaic), thermal energy production (photovoltaic), chemical energy production (hydrogen), electrical, and thermal energy storage

View on Biogas Plants	Coding/ Software	Biomass Modeling	Region	Additional Modeling
Energy system	Balmorel linear optimization (CPLEX-solver)	Annual energy potential (stable and increasing 1.3% per year)	Denmark, Germany, Finland, Norway, Sweden	Different waste to energy technologies (e.g., gasification, co-combustion) and other technologies (e.g., heater, steam turbine), all with fixed efficiencies
Energy system	Energy-PLAN	Annual chemical biomass potential	Denmark	Electrical energy production (wind, photovoltaic, wave, CHP, power plants), biogas purification
Energy system	Sifre	Annual energy potential of manure and straw	Danish municipality	-
Energy system	TIMES	Annual energy potential of degradable feedstock	Ireland	-
Regional potential	-	Electrical energy potential of manure	Northwestern Portugal	-
Regional potential	-	Sectoral biogas potential of manure (cattle, pigs, sheep, poultry)	Greece	Chronological sequence since 1970; contemplation of regional gas grid

View on Biogas Plants	Coding/ Software	Biomass Modeling	Region	Additional Modeling
Regional potential	-	Time-dependent (seasons) biogas potential of agricultural residues and municipal waste	Croatia	Residue-to-product ratios, sustainable removal rates
Regional potential	-	Methane potential of manure, grass silage, municipal waste	Finland	Maximum feasible use of regional feedstock due to 30-day HRT; optimizing GHG emissions
Regional potential	-	Municipal waste, sludge, manure, silage, and crop residues	Finland	Optimizing biogas plant placing
GHG emission	GaBi	methane yield of maize	Germany	Regional methane yield
GHG emission	GaBi	Methane yield of manure, maize silage and grass silage with different mixture ratios	-	CHP size and efficiency

View on Biogas Plants	Coding/ Software	Biomass Modeling	Region	Additional Modeling
GHG emission	SimaPro	Methane yield of maize, grass, rye silage, chicken manure	-	Demand-oriented energy production by HRT for mass flow calculation in digester
GHG emission	-	mass-specific energy of grass	-	influence of grass treatment
GHG emission	Umberto	Biogas yield of cultivated crops (maize, triticale, rye, hemp)	-	Emissions of farming, digestion, purification, and upgrading to biomethane, transportation
GHG emission	SimaPro MATLAB®	Dynamic AD model (AMOCO)	Germany	Demand-oriented energy production with dynamic AD modeling

Source: Heiker M, Kraume M, Mertins A, Wawer T, Rosenberger S. Biogas Plants in Renewable Energy Systems—A Systematic Review of Modeling Approaches of Biogas Production. Applied Sciences. 2021; 11(8):3361. https://doi.org/10.3390/app11083361.

Table 5.2. *Biogas Bottom-Up Modeling (Process Dynamics): Please Refer to the Article For Back References*

View on Biogas Plants	AD Model	Coding/Software	Feedstock	Energy Production	Region	Additional Models
AD process	ADM1: 24 species, 19 reactions	-	-	-	-	Physiochemical digester model
AD process	ADM1	SIMBA#:C#-based	-	-	-	-

View on Biogas Plants	AD Model	Coding/Software	Feedstock	Energy Production	Region	Additional Models
AD process	ADM1	SIMBA#	Pig manure	-	-	-
AD process	ADM1	AQUASIM: C++-based	-	-	-	-
AD process	ADM1	AQUASIM	Sludge	-	-	-
AD process	ADM1	AQUASIM	Water thyme	-	-	-
AD process	ADM1	MATLAB® Simulink-code: C-based S-Function	-	-	-	-
AD process	ADM1	ADMS 1.0: Python GUI and MATLAB® ADM1	-	-	-	-
AD process	ADM1	MATLAB®-code	-	-	-	-
AD process	ADM1	BioOptim	BIO WASTE	-	-	-
AD process	modified ADM1	MATLAB® Simulink	PIG MANURE AND GLYCERIN	-	-	-
AD process	2 species, 2 reactions	MATLAB® Simulink	MAIZE SILAGE	-	-	-
AD process	1 reaction	not known	MANURE	-	-	Heat flow, thermodynamics of digester
AD process	13 species, 10 reactions	MATLAB®	Fictive waste composition	-	-	Heat flow, thermodynamics of digester
AD process	ANN: one specific digester	MATLAB®	Agricultural waste (landfill)	-	-	-
AD process	ANN: 25 digesters	NeuroSolutions®	Manure, banana stem, sawdust	-	-	-

View on Biogas Plants	AD Model	Coding/Software	Feedstock	Energy Production	Region	Additional Models
AD process	ANN of ADM1	MATLAB®: ADM1 and Python: ANN	Fictive (result of ADM1)	-	-	-
Energy production	ADM1	MATLAB® Simulink	Manure	Electricity (micro gas turbine)	-	Power electronics of micro gas turbine
Energy production	ADM1	MATLAB® Simulink	Multiple	Electricity (CHP)		Thermodynamics of digester
Energy production	4 species, 4 reactions	MATLAB® Simulink	Manure	Electricity (micro gas turbine)	-	Synchronous electrical generator of micro gas turbine, gas storage, thermodynamics of digester
Energy production	1 reaction	MATLAB® Simulink	Household garbage	Electricity (CHP), heat	Domestic use profile (China)	Heat storage (water tank), electrical gas compressor, gas storage, battery (buffer)
Energy production	1 time-dependent function	MATLAB® Simulink	Manure	Electricity (CHP)	-	Gas storage
Biogas control	ADM1	BioOptim: MATLAB® Simulink	-	Electricity (CHP)	-	Digestate storage, pumps, heating system, energy sinks and sources

View on Biogas Plants	**AD Model**	**Coding/Software**	**Feedstock**	**Energy Production**	**Region**	**Additional Models**
Biogas control	ADM1	DyBiM: MATLAB® S imulink	Grass silage, cattle manure, agricultural substrates	Electricity (CHP)	Sweden	Gas storage
Biogas control	ADM1	MATLAB®	Maize silage, rye, triticale, sugar beets, potato pieces, potato peel	-	Germany	PI controller
Biogas control	13 equations, 2 reactions	FORTRAN and WinErs for GUI and automation	-	-	-	Tanks, valves, pumps
Biogas control	ADM1 simplification	not known	Not known	Electricity (CHP)	Germany (EPEX)	Gas storage
Biogas control	linear equation	HOMER®	Undifferentiated	Electricity (CHP, photovoltaic, fuel cell)	India (off-grid)	Heat storage, energy storage (battery)
Biogas control	1 species, 1 reaction	MATLAB® and Microsoft Excel®	Maize silage, grass silage, manure	Electricity (CHP), fuel (CNG)	Germany (EPEX)	Biogas to CNG upgrade plant (black box), vehicle fleet
Biogas control	None	RedSim	Fixed gas characteristics	Electricity (CHP)	Germany (spot market)	Gas storage (mass balance)

View on Biogas Plants	AD Model	Coding/Software	Feedstock	Energy Production	Region	Additional Models
Biogas control	None	IPSEpro®	Real data gas characteristics	Electricity (CHP), fuel (methane)	-	Gas storage, heat storage, tanks, gas upgrade (black box)

Source: Heiker M, Kraume M, Mertins A, Wawer T, Rosenberger S. Biogas Plants in Renewable Energy Systems—A Systematic Review of Modeling Approaches of Biogas Production. Applied Sciences. 2021; 11(8):3361. https://doi.org/10.3390/app11083361.

5.4.5.1. Anaerobic Environment

Since methanogens are inherently anaerobic, they require an oxygen-free environment. For this reason, a biogas reactor needs to be sealed. The minimal quantity of oxygen dissolved in the liquid or biomass supplied to the plant is rapidly consumed by various microorganisms, such as facultative anaerobic bacteria that can utilize oxygen for respiration, or aerobic bacteria that require oxygen.

5.4.5.2. Temperature

The rate of biochemical processes generally increases with temperature. As a rule of thumb, the rate is doubled for every 10° rise in temperature within certain limits (Q_{10} = 2). This is also the case with the biogas process. In this situation there are, however, several types or strains of bacteria involved that have adapted to the different temperatures:

- psychrophiles 0–20□.
- mesophiles 15–45□.
- thermophiles 40–65□.

Common to the bacteria is that they are very sensitive to changes in temperature. This sensitivity increases with temperature. In practice, biogas plants are run at either a mesophilic level of around 37□, where fluctuations of approx. ± 2□ are tolerated, or at a thermophilic level of around 52□, where fluctuations of only approx. ± 0.5□ are tolerated.

5.4.5.3. Acidity (pH)

The methanogens cannot survive in an acidic environment, even though they consume some organic acids as food. A pH of between 6 and 8 is the ideal range, with 7 being the preferred value. The reactor's acidity will fall within this range when the process is in balance, and because of its enormous buffer capacity, changing it will require a lot of work. Put another way, the system is incredibly reliable and steady. Because they contain more ammonium, plants that are slurry-based typically have a slightly higher pH (8–83 points).

Despite the methanogens using organic acids for some of their food intake, they cannot cope in an acidic environment. The optimum environment is a pH of between 6.5 and 8, and the preferred level is 7.2. When the process is in balance, the acidity in the reactor will be within this range and as the buffer capacity in the reactor is very large, it takes a lot to alter it. The system is, in other words, very robust and stable. Slurry-based plants often have a somewhat higher pH (8–8.3) due to a higher ammonium content.

5.4.5.4. Substrate (feedstock)

Nearly all organic matter can be decomposed anaerobically, but the degree of decomposition can be increased in various ways. Lignin is, however, indigestible.

5.4.5.5. Comminution

The finer the material, the larger the relative surface and the easier it is for the bacteria to attack the material.

5.4.5.6. Dry Matter Content

The material's dry matter content must not exceed roughly 50% for bacteria to break it down. However, for a biogas plant to stay sufficiently liquid to be pumped, it should only be between 8 and 10%. Some types of special reactors that have a direct feed line can withstand a slightly higher level.

5.4.5.7. Carbon/nitrogen (C/N) Ratio

Methanogens require a variety of macro- and micronutrients to flourish, just like any other organism. The three macronutrients that are most crucial are potassium (K), phosphorus (P), and nitrogen (N). Bacteria use nitrogen to make proteins. Since it indicates whether there is enough nitrogen available for bacteria, the nitrogen content is frequently quoted in relation to carbon. Generally, nitrogen becomes the limiting factor for bacterial growth when the

C/N ratio is greater than 30/1. However, too much nitrogen might also impede the process, so it's important to watch that level.

5.4.5.8. Stirring

There are a number of different plant types, but for the most common type – CSTR (Continuously Stirred Tank Reactors) – the biomass has to be vigorously agitated to avoid the formation of an impenetrable surface crust.

5.4.5.9. Organic Load

It is necessary to match the rate of biomass addition to the methanogen growth in the reactor and to remove organic acids at the rate at which they are generated. A CSTR reactor typically has a load of 1–6 kg COD/m^3 reactor volume/day. The process will turn acidic if more biomass is added then the bacteria can break down. Additionally, the biomass must be fed into the reactor continuously, ideally at a consistent rate and volume. For bacteria to adjust to the new conditions, any changes to the substrate must be made gradually.

5.4.6. Inhibition of the Biogas Process

When a substance inhibits bacteria, it affects them adversely without actually killing them. There are numerous ways to hinder the process, which are frequently separated into endogenous and exogenous causes. Exogenous inhibition results from external conditions, whereas endogenous inhibition results from conditions or material generated during the process itself that, in certain situations, may inhibit the process.

5.4.6.1. Nitrogen Inhibition

Ammonia (NH_3) is one of the most important endogenous inhibitors. When nitrogen-containing materials like proteins are broken down by bacteria, NH_3 is produced. Bacterial growth requires nitrogen, and NH_3 is a significant source of nitrogen. However, high concentrations of NH_3 are extremely harmful to the bacteria (Wilkie, 2001; Graunke and Wilkie, 2008; Abatzoglou, 2009).

In an aqueous solution, NH_3 is always found in an equilibrium with ammonium (NH_4^+). This equilibrium is determined by the acidity, pH, and temperature of the environment and, as ammonium is not as toxic as NH_3, this equilibrium is important:

$$NH_4^+ \longleftrightarrow NH_3 + H^+$$

A high pH causes the environment to become more toxic to bacteria by shifting the equilibrium to the right. This equilibrium will also move to the right at higher temperatures. Because of this, a thermophilic biogas process is more sensitive to NH_3 inhibition than a mesophilic process, assuming all other factors remain constant. The bacteria will already be somewhat inhibited at relatively low NH_3 concentrations. Bacteria, however, can adapt to a greater concentration if they have a longer adaptation period. This is lucky since the biomasses that are commonly used to produce biogas, like slurry, typically have NH_3 concentrations that are towards the higher end of the spectrum. A sudden increase in concentration will be difficult for the bacteria to handle, so a steady and even biomass input is crucial for the process—even more so at higher temperatures (Table 5.3).

Table 5.3. Biogas Production Using Selected Biowaste Resources

Substrate	DM%	VS% of DM	Biogas Yield per Ton Fresh Matter (m^3)	Methane Content (%)
Pig slurry	3–8	70–80	340–550	65–70
Cattle slurry	6–12	70–85	90–310	65
Poultry droppings	10–30	70–80	310–620	60
Wheat straw	85–90	90–95	200–300	50–60
Rye straw	85–90	90–95	200–300	59
Barley straw	85–90	90–95	250–300	59
Oat straw	20–25	90–95	290–310	59
Corn straw	30–40	90–95	380–460	59
Hemp	20–30	90–95	360	59
Grass	20–30	90–95	280–550	70
Elephant grass	20–30	90–95	430–560	60
Sunflower leaves	15–20	90–95	300	59
Agricultural waste	10–40	75–90	310–430	60–70
Slaughterhouse residues	35	90–95	550–650	58
Algae	20–40	90–95	420–500	63

Source: Singh AK, Pal P, Rathore SS, Sahoo UK, Sarangi PK, Prus P, Dziekański P. Sustainable Utilization of Biowaste Resources for Biogas Production to Meet Rural Bioenergy Requirements. Energies. 2023; 16(14):5409. https://doi.org/10.3390/en16145409.

5.4.6.2. Acidification – Organic Acids

Other important endogenous process inhibitors are the organic acids formed during the process. If these are not removed as soon as they are formed – which can happen during an overload – this can lead to an acidification of the process.

5.4.6.3. Antibiotics

Since both are toxic to and intended to kill microorganisms, antibiotics, and disinfectants are clear examples of exogenous causes that inhibit the process. Both compounds can be found in the slurry because they are used in livestock production to treat sick animals and maintain clean animal houses and milking parlors. However, it appears that their concentrations in the slurry are only low enough to not negatively affect the biogas plant. If the supply is continuously fed in, a slow adaptation to these substances may also occur.

At high concentrations, other substances like salts, micronutrients, and heavy metals can also impede the process. However, as was already mentioned, some of them, like vitamins for humans, are necessary for the process at low concentrations (Wilkie, 2001; Graunke and Wilkie, 2008; Abatzoglou, 2009).

5.4.7. An Alternative Energy Source

Fewer professionals in the energy sector are familiar with biogas; the majority of us are familiar with biomass. The fundamentals of biogas are discussed in this commentary, with an emphasis on its present applications and potential as a source of electricity in the future.

Anaerobic digestion of "biomass," or biodegradable organic matter, yields biogas. Employing biochemical reactions, bacteria that are present in or added to the biomass ferment it anaerobically—that is, without oxygen. Methane (60–80%), carbon dioxide (20–40%), and trace amounts of nitrogen, hydrogen sulfide, and other impurities are the components that makeup biogas. Biogas must first be cleaned to remove moisture and impurities before it can be used as an energy source for some alternatives, like compressed natural gas. The process of purifying biogas is referred to as "upgrading." "Biomethane" is the result of rendering biogas to contain at least 98% methane. Biomethane is one of the cleanest and most effective alternative energy sources; it functions exactly like natural gas, the traditional fossil fuel, and has all the same advantages and applications.

Landfills, industrial waste treatment facilities, municipal wastewater treatment plants, and agricultural sources like energy crops and manure are some of the main sources of biogas. Anaerobic digestion was previously employed by these facilities to stabilize, reduce pathogens, and reduce waste volume before disposal or land application. Biogas was frequently just flared.

These days, biogas is becoming a major alternative energy source. Using biogas to generate electricity simultaneously addresses several regulatory requirements. Greenhouse gas emissions are minimized due to the inhibition of methane release, the generation of sustainable and renewable energy, and the decrease in the amount of waste that needs to be stored and disposed of.

5.4.7.1. Uses of Biomethane

Biomethane has several benefits, such as being used for combined heat and power (CHP), boiler fuel, injection into natural gas pipelines, and vehicle fuels in the form of liquefied and compressed biomethane. Biogas can be used by CHP systems, which generate both mechanical and thermal energy, to generate electricity. The generated energy can be fed into conventional power grids or used on-site, which lowers the facility's electrical expenses. Internal combustion engines, combustion gas turbines, micro turbines, fuel cells, and steam turbines are among the CHP applications for biogas.

In other nations, particularly those in Europe, where the price of electricity is substantially greater than in the U.S. Anaerobic digestion is a commonly employed method for producing biogas. Facilities for anaerobic digestion using "energy crops," like corn, are being constructed or are in the process of being constructed in Germany, Sweden, Poland, Hungary, and Denmark. These establishments generate biogas, but in the U.S. energy crops are utilized to produce ethanol or fuel alcohol. When the same amount of energy crops are converted to alternative energy, more net energy value is produced through the production of biogas than ethanol. This is because biogas production is significantly more energy-efficient than ethanol production.

Other nations are now under pressure from the Kyoto Protocol to set goals for renewable energy and to advance the advancement of renewable energy technologies. The first country to implement feed-in-tariff (FIT) legislation supporting biogas was Germany. Utilities are required by the FIT legislation to purchase the electricity generated by biogas generators. Biogas can be used by German farmers to produce electricity that can be sold to the grid. The wider adoption of biogas in the U.S. has been impeded by certain obstacles.

These obstacles include unfavorable economic conditions, a lack of funding, challenging technical issues, and delays in obtaining air permits. The benefits throughout the entire life cycle, however, may outweigh the initial costs despite the large upfront capital investment that is required. One of the best examples of utilizing biogas to generate electricity is the Village Creek Water

Reclamation Facility in Fort Worth, Texas. This state-of-the-art municipal facility supplements its biogas production with high-strength liquid industrial wastes from six upgraded anaerobic sludge digesters, which produce low-Btu methane biogas as a byproduct.

The unprocessed, unrefined biogas is fed into two 5.2-MW combustion turbines via a dehydrator first, along with biogas from a neighboring landfill. Boilers that produce steam are fired by combining extra biogas with the waste heat from the engines in a duct burner. The facility's two 1000-horsepower blowers, which supply aeration for the activated biological treatment section, are powered by two steam turbines that are powered by the steam. When every unit is running, the facility will generate more biogas than it needs to run itself and extra electricity that it can sell to the grid or exchange for usage at other facilities (Wilkie, 2001; Graunke and Wilkie, 2008; Abatzoglou, 2009).

5.4.8. Biogas Plant Design

Distinct varieties of biogas plants exist that can handle various forms of biomass, and each has pros and cons of its own. However, only CSTRs—continuously stirred tank reactors—are used in Danish farm and community biogas plants. One benefit of the CSTR is its ability to handle biomasses with a comparatively high dry matter content.

Regular batches of biomass are fed into the reactor either continuously or semi-continuously. The biggest disadvantage of this kind of reactor is that some of the freshly added biomass is pumped out again too soon and before it is fully decomposed because some material must be pumped out to make room for the new biomass input. This is because of the constant stirring (Table 5.4).

Table 5.4. *Advantages and Disadvantages of Anaerobic Digestion*

Advantages	Disadvantages
• Energy production by producing high quality soil fertilizer	Less heat released, resulting in lower and less efficient destruction of pathogens as in aerobic composting
• No need for additional power to turn the pile of waste for the purpose of obtaining oxygen	Unsuitable for waste containing less organic matter

Advantages	Disadvantages
• Closed system allows the use of all produced gas	Requirement for waste separation to improve decommissioning efficiency
• Monitoring of greenhouse gas emissions	Pretreatment is essential
• No unwanted odor, rodents, and flies	Temperature sensitivity
• The modular construction of the plant and closed process require smaller land (footprint) areas	Post-processing is required
• Net positive environmental gains	2–4 months of start-up time
• Possible implementation on a small scale	
• Low power consumption	
• Almost complete retention of nutrients in the fertilizer	
• Possibility to store sludge for a longer period	
• Construction costs are relatively low	
• Low sludge production	
• Low nutrient demand	
• High organic removal	

Source: Zupančič M, Možic V, Može M, Cimerman F, Golobič I. Current Status and Review of Waste-to-Biogas Conversion for Selected European Countries and Worldwide. Sustainability. 2022; 14(3):1823. https://doi.org/10.3390/su14031823.

In the industrial setting, so-called filter plants—such as those that use UASBs, or Upflow Anaerobic Sludge Blankets—are occasionally employed to treat biomasses with low dry matter contents. This type has the advantage that the reactor tank does not need to be as large because the (hydraulic) retention time—the amount of time a given biomass stays in the reactor before it is pumped out again—is very short, often only a few hours or a couple of days.

A comparatively high COD load of 5–30 kg COD/m^3/day can also be handled by this type. Since the fully stirred plant type is the most widely utilized in Denmark for the breakdown of sewage sludge and agricultural residues, only this type will be covered in the following. In general, the designs of communal and on-farm plants are similar, but certain sections will clearly differ in size based on factors like the amount of biomass they are intended to manage (Wilkie, 2001; Graunke and Wilkie, 2008; Abatzoglou, 2009).

REFERENCES

1. Abatzoglou, N., & Boivin, S. (2009). A review of biogas purification processes. *Biofuels, Bioproducts and Biorefining, 3*(1), 42–71. https://doi.org/10.1002/bbb.142.
2. Biogas flows through Germany's grid big time. (2012). *Renewable Energy News Article*.
3. Cornejo, C., & Wilkie, A. C. (2010). Greenhouse gas emissions and biogas potential from livestock in Ecuador. *Energy for Sustainable Development, 14*(4), 256–266. https://doi.org/10.1016/j.esd.2010.09.002.
4. Graunke, R. E., & Wilkie, A. C. (2008). Converting food waste to biogas: Sustainable Gator dining. *Sustainability: The Journal of Record, 1*(6), 391–394. https://doi.org/10.1089/sus.2008.2906.
5. Hedlund, F. H., & Madsen, M. (2018). Incomplete understanding of biogas chemical hazards–Serious gas poisoning accident while unloading food waste at biogas plant. *Journal of Chemical Health & Safety*.
6. Le Mer, J., & Roger, P. (2001). Production, oxidation, emission, and consumption of methane by soils: A review. *European Journal of Soil Biology, 37*(1), 25–50. https://doi.org/10.1016/S1164-5563(01)01055-3.
7. Wieland, P. (2006). Biomass digestion in agriculture: A successful pathway for energy production and waste treatment in Germany. *Engineering in Life Sciences, 6*(3), 302–309. https://doi.org/10.1002/elsc.200600035.
8. Wilkie, A. C. (2007). Biogas and biofuel synergies. Paper presented at the 2007 Farm to Fuel Summit, St. Petersburg, Florida.

Chapter **6**

Geothermal Energy

Contents

INTRODUCTION

The Greek terms geo (earth) and therme (heat) are the source of the word geothermal. The heat from within the earth is therefore geothermal energy. Buildings can be heated or electricity can be produced using the steam and hot water produced inside the earth. Because heat is constantly generated inside the earth and water is constantly replenished by rainfall, geothermal energy is a renewable energy source. The heat that comes from the earth's subsurface is known as geothermal energy.

The geothermal energy is transported to the Earth's surface by steam or water. Geothermal energy can be used to produce clean electricity or used for heating and cooling, depending on its properties. But to generate electricity, high- or medium-temperature resources are required, and these are typically found close to tectonically active areas.

6.1. CONCEPT OF GEOTHERMAL ENERGY

Geothermal energy is a type of energy conversion where heat energy from within the Earth is captured and used for various purposes such as cooking, bathing, heating space, and producing electricity. Lava flows, geysers, fumaroles, hot springs, and mud pots are examples of surface phenomena that are created by heat from Earth's interior. The primary sources of heat are friction created at the margins of continental plates and the radioactive decay of uranium, thorium, and potassium in the Earth's crust and mantle. Around the world, the ensuing yearly low-grade heat flow to the surface ranges from 50 to 70 milliwatts (mW) per square meter. On the other hand, the Earth's surface receives 342 watts per square meter of annual solar radiation. Anywhere on Earth's surface, geothermal heat energy can be extracted and used for human purposes. Roughly three times the world's annual energy consumption of all forms combined, or 4.5×10^6 exajoules, or approximately 1.4×10^6 terawatt-years, are estimated to be recoverable and used on the surface (Figure 6.1).

Geothermal energy yields varying amounts depending on extraction technique and depth. In the upper part of the lithosphere, the average global increase in temperature of rocks and other subsurface materials is 20–30□ (36–54°F) per kilometer (0–0.6 mile), and this rate of increase is significantly higher in the majority of Earth's known geothermal areas. Typically, a fluid (or steam) is needed for heat extraction to raise the energy to the surface. It can be difficult to find and develop geothermal resources.

Figure 6.1. *The Nesjavellir geothermal power plant, Iceland.*

Source: By Gretar Ívarsson – Edited by Fir0002 – Gretar Ívarsson, geologist at Nesjavellir, Public Domain, https://commons.wikimedia.org/w/index.php?curid=2523755.

This is particularly true for the resources that require high temperatures to produce electricity. These kinds of resources are usually found in regions of the world where there has been recent volcanic activity, along plate boundaries, or inside crustal hot spots. Earth has a constant supply of heat, but because heated fluids and steam can be extracted at a rate faster than they can be replenished, sustainable resource management is required (Office of Energy Efficiency and Renewable Energy, 2013; U.S. Department of Energy, 2014).

6.1.1. Uses of Geothermal Energy

Humans have utilized the Earth's inherent geothermal energy for thousands of years. Geothermal energy use dates back to Ancient China and Egypt, as well as the Greek and Roman eras. This versatile energy source can be applied in various ways, and as people and governments began seeking renewable energy options in the early 21st century, geothermal energy gained significant attention among alternative energy advocates. The most common application of geothermal energy is direct use, which involves heating, cooking, and bathing with hot water and steam from within the Earth. The earliest use is direct application; Greeks and Romans, for instance, believed that naturally heated water deposits had therapeutic properties and used geothermal springs for bathing. This application is widespread in more than 70 nations worldwide, and it is particularly prevalent in geologically active areas like Iceland, where geothermal energy is abundant (Figure 6.2).

Figure 6.2. *Geothermal resources are usually found along land areas where earthquakes and volcanoes occur. The area where the energy is concentrated is called the ring of fire which covers the Pacific Ocean.*

Source: By Ytrottier – Own work, CC BY-SA 3.0, https://commons.wikimedia.org/w/index.php?curid=7278148.

Geothermal power plants, which use geothermal energy to generate electricity, are another application for geothermal energy. These facilities have the benefit of being able to produce energy in one place and move it to another, giving residents of areas lacking geothermal energy sources access to geothermal power. Depending on the technology employed, the level of maintenance, and the location of the power source, geothermal power plants can have extremely high-efficiency rates.

A heat pump is an additional geothermal energy application option. Geothermal energy is used in conjunction with a heat pump to both push and pull heat into and out of buildings. This method controls the climate, frequently for very little money. It doesn't need access to deep geothermal deposits; instead, it depends on the Earth's inherent warmth near the surface. To distinguish this use of geothermal energy from the deep geothermal energy deposits utilized in geothermal power plants and direct use systems, it may be referred to as geo-exchange.

Like any other alternative energy source, this one has benefits and drawbacks. People use it for cooking, heating, powering cars, producing electricity, and other purposes. Many proponents of alternative energy have hypothesized that rather than depending solely on one technology, people should combine energy sources and technologies to meet their needs in a happy medium.

6.1.1.1. Direct Uses

Direct use of heated groundwater without the need for specialized equipment is likely the most popular set of applications. Low-temperature geothermal resources, which range from roughly 50 to 150□ (122 to 302°F), are used in all direct-use applications. Such low-temperature geothermal steam and water have been used to heat entire districts, encompassing multiple buildings heated by a single central supply source, as well as individual buildings. Furthermore, geothermal resources have been used to heat a large number of swimming pools, greenhouses, aquaculture ponds, and balneological (therapeutic) facilities at spas worldwide. Cooking, industrial applications (like drying fruit, vegetables, and wood), pasteurizing milk, and massive snow melting are some additional direct uses of geothermal energy. Hot water is frequently used in the heating system directly for many of those purposes, or it can be used in conjunction with a heat exchanger, which transfers heat when the fluid contains problematic minerals and gases like hydrogen sulfide (Figure 6.3).

Figure 6.3. *A new battery cell design could help us tap into the vast stores of geothermal energy beneath our feet.*

Source: By Jack Catalano – Own work, CC BY 4.0, https://commons.wikimedia.org/w/index.php?curid=84570597.

6.1.1.2. Geothermal Heat Pumps

Buildings can be heated or cooled in the winter and summer by using geothermal heat pumps (GHPs), which take advantage of the generally stable, moderate temperature conditions that exist within the first 300 meters (1000 feet) of the surface. Rocks and groundwater are found in that region of the lithosphere, where temperatures range from 5 to 30□ (41 to 86°F). The majority of GHPs are found at shallower depths, such as within 6 meters (roughly 20 feet) of the Earth's surface, where the ground temperature remains relatively constant at 10 to 16□ (50 to 60°F). As a result, during the colder months of the year, when the air temperature drops below that of the ground, that heat can be utilized to help warm buildings. In a similar vein, warm air can be extracted from a building and moved underground during the warmer months of the year, where it loses most of its heat before returning.

A pump and a heat exchanger, which is a subterranean loop of pipes, make up a GHP system. Using a fluid that circulates through the pipes, the heat exchanger transfers heat energy between the ground and the air at the surface. Water or a solution of water and antifreeze is frequently used as the fluid. Warm air transfers heat to the heat exchanger and then into the fluid during the warmer months. The heat is distributed to the rocks, soil, and groundwater as it passes through the pipes. In the winter, the pump is turned the other way around. The fluid's temperature rises due to heat energy retained in the comparatively warm ground. The heat pump uses the fluid to transmit this energy, warming the building's interior air (Figure 6.4).

Figure 6.4. *Geothermal sources.*

Source: By Stepheng3 – Own work, CC0, https://commons.wikimedia.org/w/index.php?curid=19087599.

Compared to more traditional heating and air conditioning systems, GHPs have a number of advantages. They are less polluting and extremely efficient, consuming 25–50% less electricity than equivalent conventional heating and cooling systems. When compared to air-source heat pumps, which transfer heat between indoor and outdoor air, the energy-saving benefits of GHPs can result in a reduction in greenhouse gas emissions of up to 44%. Furthermore, GHPs can reduce greenhouse gas emissions by up to 72% when combined with conventional air conditioning systems and electric resistance heating systems, which turn electricity into heat.

6.1.1.3. Electric Power Generation

Electricity can be produced from geothermal energy depending on temperature and fluid (steam) flow. There are three ways that geothermal power plants can generate electricity. All three use steam to power electrical generators by controlling its behavior, despite having different designs. Geothermal power is regarded as a renewable energy source since it condenses excess water vapor at the end of each process and returns it to the ground to be heated again later. Certain geothermal power plants merely gather the earth's rising steam. The heated water vapor is directed straight into a turbine in these "dry steam" operations, which powers an electrical generator.

Other power plants, which are based on the flash steam and cycle designs, begin the process of producing electricity by heating water (also known as "wet steam") and steam that is extracted from the ground (Figure 6.5).

Figure 6.5. *Coal fired power plant.*

Source: By Morgre – Own work, CC BY-SA 3.0, https://commons.wikimedia.org/w/index.php?curid=19858375.

In flash steam power plants, high-temperature water that has been pressurized is extracted from below the surface and placed in containers at the surface known as flash tanks. When the pressure suddenly drops, the liquid water vaporizes and turns into steam. The turbine-generator set is then powered by the steam. On the other hand, steam driven off a secondary working fluid (like ammonia and hydrocarbons) contained within a closed loop of pipes powers the turbine-generator set in binary-cycle power plants. Through a separate set of pipes, geothermally heated water is drawn up for this process, and a heat exchanger transfers most of the energy stored in the heated water to the working fluid. After that, the working fluid evaporates. The working fluid vapor is condensed and returned to the heat exchanger after passing through the turbine.

Electrical power usually requires water heated above 175□ (347°F) to be economical. In geothermal plants using the Organic Rankine Cycle (ORC), a special type of binary-cycle technology that utilizes lower-temperature heat sources (such as biomass combustion and industrial waste heat), water temperatures as low as 85–90□ (185–194°F) may be used (Office of Energy Efficiency and Renewable Energy, 2013; U.S. Department of Energy, 2014).

6.1.2. Extraction of Geothermal Energy

High-temperature gradients are ideal for the production of geothermal energy (Figure 6.6). These gradients can be found in places that have recently experienced volcanism, in regions that are situated near plate boundaries (like the Pacific Ring of Fire), or in regions that are characterized by hot spots or thin crust, like the Hawaiian Islands and Yellowstone National Park. Geothermal reservoirs in those areas need to have an impermeable caprock to keep heat from escaping, a heat source, sufficient water recharge, and a reservoir with sufficient permeability or faults that permit fluids to rise near the surface. Furthermore, these reservoirs have to be economically accessible (i.e., within drillable range).

Drilling wells up to 9100 meters (30,000 feet) allows for the extraction of heated fluid from a geothermal resource, which is then pumped out or naturally extracted through artesian flow, in which the water's weight pushes it to the surface. After that, water, and steam are piped to the power plant to produce electricity, or they can be used for heating and cooling purposes through insulated pipelines that can be buried or installed aboveground. To reduce heat loss in the steam, pipelines used in electric power plants are typically restricted to a distance of approximately 1 mile or 1.6 kilometers. Nonetheless, depending on the flow rate, direct-use pipelines spanning tens of kilometers have been installed with less than a 2–5□ (3.6–9°F) temperature loss. Because building lengthy pipelines costs money, the most economical facilities are situated close to the geothermal resource. When it comes to the production of electricity, expenses can be reduced by situating the facility close to electrical transmission lines that carry the electricity to the market.

Figure 6.6. *Geothermal energy: developing countries are showing a trend.*

Source: By Mike Schiraldi at the English Wikipedia, CC BY-SA 3.0, https://commons.wikimedia.org/w/index.php?curid=6901185.

6.1.3. Exhaustion of Geothermal Energy

If the rate of heat extraction outpaces the rate of natural heat recharge, geothermal resources may run out. Geothermal resources can typically be used for 20 to 30 years; however, over time, the energy output may decline, making further development unfeasible. Conversely, since the early 1900s, geothermal electricity has been continuously generated at the Geysers and from the Larderello geothermal field since 1960. Despite the fact that both of those fields have declined, this issue has been somewhat resolved by recharging the water supply and digging new wells. Electrical capacity at the Geysers decreased from 1800 MW to roughly 1000 MW, but 200 MW of capacity was recovered by unifying the field under one operator and building pipelines to transport wastewater for reservoir replenishment. Certain projects, like the geothermal heating system at Oregon Institute of Technology, have been in operation since the 1950s without affecting production, and projects like the district heating system in Reykjavík have been in operation since the 1930s with little change in output. Geothermal resources can therefore be sustained for a long time with appropriate management, and they can even recuperate from a temporary suspension of use (Figure 6.7).

Figure 6.7. *A geothermal power station in Poland that creates electricity from heat generated in Earth's interior.*

Source: By Morgre – Own work, CC BY-SA 3.0, https://commons.wikimedia.org/w/index.php?curid=19858375.

6.1.4. Environmental Effects and Economic Costs

Changes in land use related to exploration and plant construction, noise, and visual pollution, water, and gas discharge, the creation of unpleasant odors, and soil subsidence are some of the environmental effects of geothermal development and power generation. However, the majority of those effects are negligible with the technology available today, meaning that geothermal uses will only have a minor environmental impact. For instance, Klamath Falls, Oregon, has about 600 residential space heating geothermal wells. Additionally, the city has invested in a downtown snow-melting system and a district heating system that serves local businesses with heating. However, none of the geothermal energy delivery and supply systems are visible from the town (Table 6.1).

Table 6.1. *Major Geothermal Plants in Germany*

Selected Operating Parameters of Major Geothermal Installations in Germany. TVD: True Vertical Depth.									
Location	**Installed capacity (MW)**		**Production (GWh/a)**		**Wellhead temperature (□)**	**Production rate (L/s)**	**Depth (m TVD)**	**Stratigraphy**	**Reportingyear**
	Power	**Heat**	**Power**	**Heat**					
North German Basin									
Neubrandenburg *	-	-	-	-	65–78	11–28	1268	Rhaetian	2011
Neustadt-Glewe	-	4.0	-	17.6	97	11–35	2450	Rhaetian	2012
Waren	-	1.3	-	3.1	63	17	1565	Rhaetian	2012
Upper Rhine Graben									
Bruchsal	0.4	-	1.2	-	123	24	2542	Middle Bunter	2013
Insheim	4.3	-	14.2	-	165	65	3800	Basement	2013
Landau (Palatinate)	3.0	5.0	13.2	3.0	159	40	3291	Bunter, Basement	2012
South German Molasse Basin									
Aschheim	-	9.8	-	48.7	85.4	39–75	2630	Malm	2013
Dürrnhaar **	5.5	-	-	-	*ca.* 135	*ca.* 130	4114	Malm	-
Erding	-	10.2	-	37.6	62–63	6–36	2359	Malm	2013
Garching	-	8.0	-	10.2	74	100	2226.3	Malm	2012

Selected Operating Parameters of Major Geothermal Installations in Germany. TVD: True Vertical Depth.

Location	Installed capacity (MW)		Production (GWh/a)		Wellhead temperature (□)	Production rate (L/s)	Depth (m TVD)	Stratigraphy	Reportingyear
	Power	Heat	Power	Heat					
Kirchstockach **	5.5	-	-	-	138.8	*ca.* 120	3881.6	Malm	-
München (Riem)	-	12.0	-	48.0	94.5	35–85	2747	Malm	2011
Oberhaching	-	38.0	-	48.8	127.5	137	3755.2	Malm	2013
Poing	-	9.0	-	37.0	76.2	100	3049	Malm	2013
Pullach ***	-	11.5	-	45.7	104 and 80	55 and 23	3505	Malm	2013
Sauerlach **,***	5.0	4.0	-	-	*ca.* 140	110	4480	Malm	-
Simbach-Braunau	-	7.0	-	48.0	80.5	61.1	1941	Malm	2013
Straubing	-	2.1	-	2.9	36.5	17.5	824.8	Malm	2011
Traunreut **	-	7.0	-	-	*ca.* 108	50	4645.8	Malm	-
Unterföhring	-	9.5	-	29.0	86	*ca.* 75	1986	Malm	2012
Unterhaching	3.4	38.0	6.9	107.6	123.3	120	3590	Malm	2013
Unterschleißheim	-	8.0	-	40.1	78	65–93.3	1960	Malm	2013
Waldkraiburg	-	16.4	-	2.8	109	20	2718	Malm	2012

Selected Operating Parameters of Major Geothermal Installations in Germany. TVD: True Vertical Depth.

Location	**Installed capacity (MW)**		**Production (GWh/a)**		**Wellhead temperature (□)**	**Production rate (L/s)**	**Depth (m TVD)**	**Stratigraphy**	**Reportingyear**
	Power	**Heat**	**Power**	**Heat**					
Total	27.1	200.7	35.5	530.1	-	-	-	-	-

* heat storage facility; ** recent commissioning; and *** triplet.

Source: Agemar T, Weber J, Schulz R. Deep Geothermal Energy Production in Germany. Energies. 2014; 7(7):4397-4416. https://doi.org/10.3390/en7074397.

Furthermore, because GHPs utilize shallow geothermal resources that are located within 100 meters (or 330 feet) of the surface, they have very little environmental impact. GHPs only slightly alter the temperature of the earth's rocks and soil or groundwater. When using closed-loop systems, the ground temperature around vertical boreholes is slightly raised or lowered; the direction of this temperature change depends on whether heating or cooling dominates the system, which would be the case in colder climates. The ground temperatures will not change as long as the heating and cooling loads are balanced. Similarly, open-loop systems that use lake or groundwater would hardly affect the temperature, particularly in areas with high groundwater flows. When comparing geothermal energy's advantages to those of other renewable energy sources, the primary one is that geothermal energy's base load is available around the clock, seven days a week, while solar and wind energy are only available for roughly one-third of the time.

Furthermore, geothermal energy can be economically competitive with other energy sources like coal, with costs per kilowatt-hour ranging from 5 to 10 cents. The primary drawback of developing geothermal energy is the high upfront costs associated with building the necessary infrastructure and facilities, as well as the high risk of resource verification. Low-permeability rocks frequently contain geothermal resources, and exploration efforts frequently drill "dry" holes—that is, holes that release steam in quantities that are not profitable to extract. But once the resource is established, the yearly cost of fuel—that is, steam, and hot water—is cheap and doesn't usually increase in price (Office of Energy Efficiency and Renewable Energy, 2013; U.S. Department of Energy, 2014).

6.1.5. Advantages and Disadvantages of Geothermal Energy

Geothermal comes from the Greek Word Geo which means earth and thermo which means heat. So, geothermal energy means energy or power extracted from beneath the earth. The energy inside the earth was formed by the decay of minerals and forests several years ago. Traditionally, it was used for bathing and heating purposes but today it is also used for generating electricity. Geothermal energy is called a renewable source of energy because heat is continuously produced inside the earth.

This heat is brought to the near-surface by thermal conduction and by intrusion into the earth's crust of molten magma originating from great depth. As groundwater is heated, geothermal energy is produced in the form of hot water and steam.

Geothermal energy is produced inside the earth's surface. The earth's layer consists of the innermost layer called the iron Core which itself has two layers: a solid iron core and an outer core made of hard rock, called magma mantle which surrounds the core and the outermost layer called the crust which forms oceans and continents.

When magma approaches the surface of the earth, that is, heats the groundwater that becomes trapped in porous rocks due to the crust. They could also move along rock surfaces that are fractured and faulted. The two components of these hydrothermal resources are heat (thermal) and water (hydro). These naturally occurring hydrothermal resources are referred to as geothermal reservoirs. These days, geothermal reservoirs are found using a variety of instruments and methods.

6.1.5.1. Advantages of Geothermal Energy

6.1.5.1.1. Significant Cost Saving

Geothermal energy generally involves low running costs since it saves 80% costs over fossil fuels and no fuel is used to generate the power. Since, no fuel is required so costs for purchasing, transporting, and cleaning up plants are quite low.

6.1.5.1.2. Reduce Reliance on Fossil Fuels

Dependence on fossil fuels decreases with the increase in the use of geothermal energy. With the sky-rocketing prices of oil, many countries are pushing companies to adopt these clean sources of energy. Burning of fossil fuels releases greenhouse gases which are responsible for global warming (Figure 6.8).

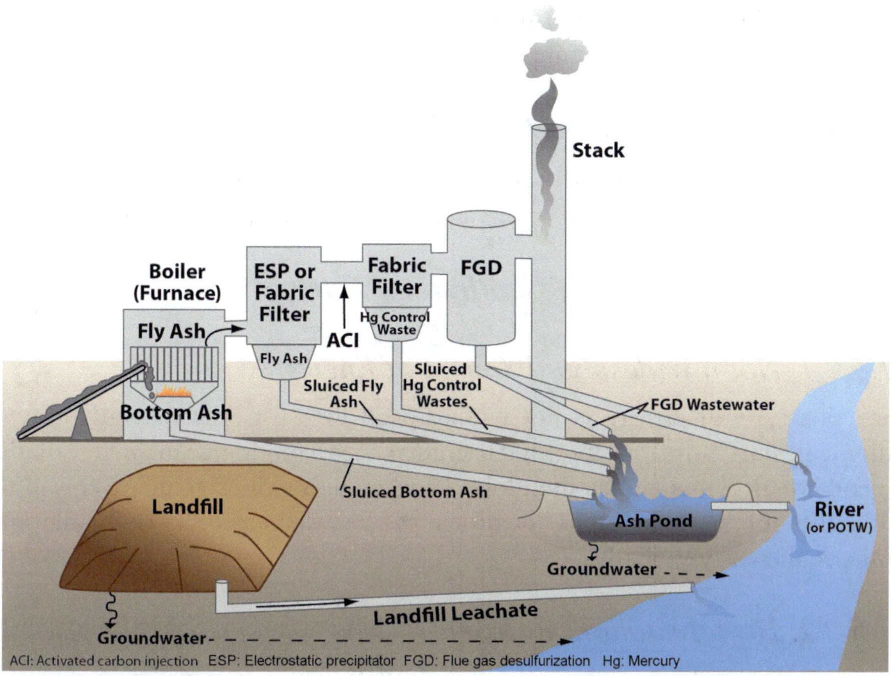

Figure 6.8. *Oil, gas, and other fossil fuels come with grave consequences for our health and our future.*

Source: By U.S. Environmental Protection Agency, Washington, D.C. – U.S. Environmental Protection Agency (EPA). "Steam Electric Power Generating Effluent Guidelines—2015 Final Rule," Public Domain, https://commons.wikimedia.org/w/index.php?curid=78669933.

6.1.5.1.3. No Pollution

One of the primary benefits of utilizing geothermal energy is its zero-pollution nature, which contributes to maintaining a clean environment. As a renewable energy source, geothermal energy has contributed to a decrease in pollution and global warming. Furthermore, because the gases released by geothermal systems are not very harmful to the environment, they come from deep within the earth and do not cause pollution.

6.1.5.1.4. Direct Use

Since ancient times, people have been using this source of energy for taking baths, heating homes, and preparing food and today this is also used for direct heating of homes and offices. This makes geothermal energy cheaper and affordable. Although the initial investment is quite steep in the long run with huge cost savings it proves quite useful.

6.1.5.1.5. Job Creation and Economic Benefits

Governments of various countries are investing hugely in the creation of geothermal energy which on the other hand has created more jobs for the local people.

6.1.5.2. Disadvantages of Geothermal Energy

6.1.5.2.1. Not a Widespread Source of Energy

Since this type of energy is not widely used therefore the unavailability of equipment, staff, infrastructure, and training poses a hindrance to the installation of geothermal plants across the globe. Not enough skilled manpower and the availability of suitable build locations pose serious problems in adopting geothermal energy globally.

6.1.5.2.2. High Installation Costs

Getting geothermal energy, requires the installation of power plants, to get steam from deep within the earth and this requires huge one-time investments and requires a certified installer and skilled staff needs to be recruited and relocated to plant locations. Moreover, electricity towers, and stations need to be set up to move the power from the geothermal plant to the consumer (Figure 6.9).

Figure 6.9. *Krafla, a geothermal power station in Iceland.*

Source: https://upload.wikimedia.org/wikipedia/commons/3/3b/Krafla_geothermal_power_station_wiki.jpg.

6.1.5.2.3. Can Run out of Steam

Geothermal sites can run out of steam over a period due to drop in temperature or if too much water is injected to cool the rocks and this may result huge loss for the companies which have invested heavily in these plants. Due to this factor, companies have to do extensive initial research before setting up the plant.

6.1.5.2.4. Suited to Particular Region

It is only suitable for regions where temperatures the earth are quite low and can produce steam over a long period of time. For this great research is required which is done by the companies before setting up the plant and this initial cost runs up the bill in setting up the geothermal power plant. Some of these regions are near hilly areas or high up in mountains.

6.1.5.2.5. May Release Harmful Gases

Geothermal sites may contain some poisonous gases and they can escape deep within the earth, through the holes drilled by the constructors. The geothermal plant must therefore be capable enough to contain these harmful and toxic gases (Figure 6.10).

Figure 6.10. *Poisonous gas emission.*

Source: By Yomangani – Own work, Public Domain, https://commons.wikimedia.org/w/index.php?curid=1396466.

6.2. GEOTHERMAL POWER PLANT

Geothermal power plants are used to generate electricity by the use of geothermal energy (the Earth's internal thermal energy). They essentially work the same as a coal or nuclear power plant, the main difference being the heat source. With geothermal, the Earth's heat replaces the boiler of a coal plant or the reactor of a nuclear plant.

Hot water or steam is extracted from the Earth through a series of wells and feeds the power plant. In most geothermal plants the water pulled up from the ground is returned back to the subsurface. The rate of water used is often larger than the rate of water returned, so make-up water supplies are generally needed.

In 1904, the first geothermal power plant was built in Tuscany, Italy. At this site, steam was already erupting through the Earth's crust, which made it easy to harness and transform this energy into electricity. Before this, energy from geothermal reservoirs was used as a heat source, though it was not stored for later use (Figure 6.11).

Figure 6.11. *Matsukawa geothermal power station, the first commercial geothermal power station.*

Source: https://upload.wikimedia.org/wikipedia/commons/thumb/a/a0/MATSUKAWA_Geothermal_power_station_Iwate%2CJAPAN.jpg/1200px-MATSUKAWA_Geothermal_power_station_Iwate%2CJAPAN.jpg.

The steam is typically reintroduced into the system to receive additional heating from the Earth after it cools and recondenses back into liquid water. To obtain a water supply with sufficient pressure to operate this kind of plant, it is typically required to drill down to a depth of one to two miles (1,6 and 3,2 km). Steam that escapes the earth's crust is captured and used to drive a turbine in another popular kind of geothermal power plant. Drilling is not necessary for these plants, also known as dry steam plants, as steam—often in the form of geysers—erupts naturally. Geothermal power plants can also be derived from binary cycle power plants.

These plants use hot water from the Earth's surface to turn another liquid into steam. A geothermal power plant typically produces very little pollution. Water is the primary byproduct of most of them, and it is frequently collected and recycled for use at a later time. To reduce the amount of pollution the plant produces, contaminants, and other chemicals present in the water on Earth's surface can be removed before the water is released into the environment. These plants frequently produce less than 5% of the pollution produced by fossil fuel-burning plants (Office of Energy Efficiency and Renewable Energy, 2013; U.S. Department of Energy, 2014).

6.2.1. Types of Geothermal Power Plants

Geothermal power plants come in three varieties: binary cycle plants, flash steam plants, and dry steam plants. Geothermal power plants generate electricity by using water that is naturally heated by the earth. We refer to this as geothermal water.

The turbine blades of the plants are turned by this geothermal water. Every kind of power plant uses a slightly different method to achieve this. A geothermal power plant is referred to as a "dry steam plant" when it uses steam that is extracted directly from the earth. Drilling holes into the rock until they reach the reservoir is a common practice in dry steam geothermal power plants. After that, steam that has reached at least 302°F (150□) is raised to the surface. The steam turns the turbine blades once it gets to the turbines. As a result, electricity is produced (Table 6.2).

Table 6.2. *A Summary, by Country, of the Installed Thermal Capacity (MWt), Annual Energy Use (TJ/yr and GWh/yr) and the Capacity Factor to the End of 2009*

Summary of Direct-Use Data Worldwide, 2010				
Country	**Capacity (MWt)**	**Annual Use (TJ/yr)**	**Annual Use (GWh/yr)**	**Capacity Factor**
Albania	11.48	40.46	11.2	0.11
Algeria	66.84	2098.68	583.0	1.00
Argentina	307.47	3906.74	1085.3	0.40
Armenia	1	15	4.2	0.48
Australia	33.33	235.1	65.3	0.22
Austria	662.85	3727.7	1035.6	0.18
Belarus	4.5	44.43	12.3	0.31
Belgium	117.9	546.97	151.9	0.15
Bosnia and Herzegovina	21.696	255.36	70.9	0.37
Brazil	360.1	6622.4	1839.7	0.58
Bulgaria	98.3	1370.12	380.6	0.44
Canada	1126	8,873	2464.9	0.25
Caribbean Islands	0.103	2.775	0.8	0.85
Chile	9.11	131.82	36.6	0.46
China	8898	7,5348.3	20,931.8	0.27
Columbia	14.4	287	79.7	0.63
Costa Rica	1	21	5.8	0.67
Croatia	67.48	468.89	130.3	0.22
Czech Republic	216.5	1,290	358.4	0.19
Denmark	200	2,500	694.5	0.40
Ecuador	5.157	102.401	28.4	0.63
Egypt	1	15	4.2	0.48
El Salvador	2	40	11.1	0.63

Summary of Direct-Use Data Worldwide, 2010				
Country	**Capacity (MWt)**	**Annual Use (TJ/yr)**	**Annual Use (GWh/yr)**	**Capacity Factor**
Estonia	63	356	98.9	0.18
Ethiopia	2.2	41.6	11.6	0.60
Finland	994	7966	2213.0	0.25
France	1345	12,929	3591.7	0.30
Georgia	26.51	689.34	191.5	0.82
Germany	2485.4	12,764.5	3546.0	0.16
Greece	134.6	937.8	260.5	0.22
Guatemala	2.31	56.46	15.7	0.78
Honduras	1.933	45	12.5	0.74
Hungary	654.6	9767	2713.3	0.47
Iceland	1826	24,361	6767.5	0.42
India	265	2,545	707.0	0.30
Indonesia	2.3	42.6	11.8	0.59
Iran	41.608	1,064.18	295.6	0.81
Ireland	134.45	691.91	192.2	0.16
Israel	82.4	2193	609.2	0.84
Italy	867	9941	2761.6	0.36
Japan	2,099.53	25,697.94	7138.9	0.39
Jordan	153.3	1540	427.8	0.32
Kenya	16	126.624	35.2	0.25
Korea (South)	229.3	1954.65	543.0	0.27
Latvia	1.63	31.81	8.8	0.62
Lithuania	47.6	411.52	114.3	0.27
Macedonia	47.18	601.41	167.1	0.40
Mexico	155.82	4,022.8	1117.5	0.82
Mongolia	6.8	213.2	59.2	0.99
Morocco	5.02	79.14	22.0	0.50
Nepal	2.717	73.743	20.5	0.86

Summary of Direct-Use Data Worldwide, 2010				
Country	**Capacity (MWt)**	**Annual Use (TJ/yr)**	**Annual Use (GWh/yr)**	**Capacity Factor**
Netherlands	1410.26	10,699.4	2972.3	0.24
New Zealand	393.22	9552	2653.5	0.77
Norway	1000	10,800	3000.2	0.34
Papua New Guinea	0.1	1	0.3	0.32
Peru	2.4	49	13.6	0.65
Philippines	1.67	12.65	3.5	0.24
Poland	281.05	1501.1	417.0	0.17
Portugal	28.1	386.4	107.3	0.44
Romania	153.24	1,265.43	351.5	0.26
Russia	308.2	6143.5	1706.7	0.63
Serbia	100.8	1410	391.7	0.44
Slovak Republic	132.2	3067.2	852.1	0.74
Slovenia	115.6	1,015.1	282.0	0.28
South Africa	6.01	114.75	31.9	0.61
Spain	141.04	684.05	190.0	0.15
Sweden	4460	45,301	12,584.6	0.32
Switzerland	1060.9	7714.6	2143.1	0.23
Tajikistan	2.93	55.4	15.4	0.60
Thailand	2.54	79.1	22.0	0.99
Tunisia	43.8	364	101.1	0.26
Turkey	2084	36,885.9	10,246.9	0.56
Ukraine	10.9	118.8	33.0	0.35
United Kingdom	186.62	849.74	236.1	0.14

Summary of Direct-Use Data Worldwide, 2010				
Country	**Capacity (MWt)**	**Annual Use (TJ/yr)**	**Annual Use (GWh/yr)**	**Capacity Factor**
United States	12,611.46	56,551.8	15,710.1	0.14
Venezuela	0.7	14	3.9	0.63
Vietnam	31.2	92.33	25.6	0.09
Yemen	1	15	4.2	0.48
Total	**48,493**	**423,830**	**117,740**	**0.28**

Source: Lund JW. Direct Utilization of Geothermal Energy. Energies. 2010; 3(8):1443-1471. https://doi.org/10.3390/en3081443.

High pressure water is pumped out of the reservoir in flash steam geothermal power plants. Despite the water's temperature being well above the boiling point, the pressure maintains the liquid state of the substance. When the water reaches the surface, the pressure is released, causing it to burst into steam at temperatures higher than 360°F (182□). The turbine blades are then turned by the steam. Both the water that condenses after becoming steam and the water that does not turn to steam are pumped back into the reservoir. These three kinds of geothermal power plants have one thing in common; they utilize steam turbines to produce electricity. This concept is pretty much comparable to other thermal power plants that use alternative sources of energy other than geothermal.

6.2.1.1. Dry Steam Power Plants

These plants use the naturally occurring dry steam from the earth. Following its journey from the production well to the surface and through a turbine, the steam condenses and is reinjected into the Earth after providing energy to the turbine. These are the most ancient kinds of geothermal power plants; the first one was constructed in Italy in 1904. This kind of power plant can only be used in areas where the subsurface temperature is relatively high because it needs the highest temperatures; however, it needs the least amount of fluid flow.

A shaft attached to a generator is turned by the turbine, which is turned by the steam. When the generator turns, the energy is transformed into electricity, which is then sent to homes, businesses, and other establishments via power lines and a power grid. The used steam travels to the condenser, where it is transformed into water and returned via the injection well to the hot water reservoir, where the cycle is restarted.

The traditional type of geothermal power plant is the dry steam power plant. In Larderello, Italy, the first dry steam power plant was established in 1904. Within the U.S., This kind of geothermal power generation is limited to California's high volcanic mountain regions.

6.2.1.2. Flash Steam Power Plants

Because there isn't much high-quality steam that occurs naturally, these kinds are the most prevalent. Water used in this method must be at least 180□ and must flow upward through the well under its own pressure. This temperature is lower than that of dry steam plants. A portion of the water flashes to steam as the pressure drops, and this steam passes through the turbine section. In addition to being recycled back down into the well, the water that did not turn into steam can also be used for heating. These systems are more expensive because they have more intricate components, but they are still competitive with traditional power sources.

Flash steam power plants are the most common types of geothermal power plants in the modern world. The Wairakei Power Station, built in 1958 in New Zealand, was the first geothermal power plant that utilized flash steam.

6.2.1.3. Binary Cycle Power Plants

Binary power plants are expected to be the most commonly used type of geothermal power plant in the future, as locations outside of the known hot spots begin to use geothermal energy. This is because binary cycle plants can make use of lower temperature water than the other two types of plants. They use a secondary loop (hence the name "binary") which contains a fluid with a low boiling point, such as pentane or butane. The water from the well flows through a heat exchanger which transfers its heat to this fluid, which vaporizes due to its low boiling point. It is then passed through a turbine, accomplishing the same task as steam.

This geothermal power plant is advantageous compared to the flash steam and dry steam power plants because it requires slightly cooler water (as low as 57□ (135°F) to heat a separate fluid (binary fluid) that has a lower boiling point. The power plant enables cooler geothermal reservoirs to be utilized than is necessary for the flash steam and dry steam power plants. We have learned that the flash steam and dry steam use water at temperatures higher than 182□ (455K; 360°F), which is pumped up under extremely high pressure to the electricity generation plant at the surface.

In binary cycle power plants, on the other hand, the slightly cooler water is permitted to return to the reservoir after hot water from the hot water reservoir is pumped up through the production well using pumps. Through the heat exchanger, a different fluid with a lower boiling point—known as the binary fluid—typically butane or pentane hydrocarbon is pumped up at a significantly high pressure. At the heat exchanger, the binary fluid is vaporized and directed to turn a turbine, which turns a shaft connected to a generator and electricity is generated. The vapor used to turn the turbine is then converted to water by cold air radiators and allowed to go back to the reservoir through the injection well.

To put it briefly, binary cycle power plants enable people to extract geothermal energy from hot water reservoirs that flash steam—something that would not be possible with dry steam plants. The efficiency rate of the binary cycle power plants is only 10–13%, though. It is acknowledged that Russia established a viable binary cycle power project in 1967.

Low thermal efficiency rates do affect geothermal electricity production the most. Still, there are a lot of applications for the warm water, exhaust heat, and byproducts. The question that most people have is how to transport geothermal electricity.

There are many similarities between the way that geothermal electricity is transported and other power sources like wind and solar energy. To lower losses and the amount of power transferred to the electrical grid, voltage is typically increased. Long-distance power transportation necessitates a highly insulated piping system, which raises the initial cost of producing geothermal electricity.

6.3. ENVIRONMENTAL IMPACTS OF GEOTHERMAL ENERGY

The environmental impact of geothermal energy is minimal, especially compared to fossil fuel power plants. When sited and constructed carefully, geothermal power plants can be a reliable source of renewable and environmentally-friendly electricity.

Hydrothermal plants, the most developed type of geothermal power plant, are found close to geologic "hot spots," which are regions of hot molten rock that are near the Earth's crust and produce hot water. In other areas, more access to geothermal energy may be possible through enhanced geothermal systems, also known as hot dry rock geothermal, which entail drilling beneath the surface of the Earth to reach deeper geothermal resources.

Geothermal plants also differ in terms of the technology they use to convert the resource to electricity (direct steam, flash, or binary) and the type of cooling technology they use (water-cooled and air-cooled). Environmental impacts will differ depending on the conversion and cooling technology used.

6.3.1. Water Quality and Use

Water consumption and quality may be impacted by geothermal power plants. High concentrations of sulfur, salt, and other minerals are frequently found in hot water that is pumped from subterranean reservoirs. Closed-loop water systems, found in the majority of geothermal facilities, involve pumping extracted water back into the geothermal reservoir after it has been used to produce heat or electricity. These systems contain water inside steel well casings that are affixed to the surrounding rock. In the US, there haven't been any documented instances of geothermal site water contamination. Geothermal plants also use water for re-injection and cooling. Cooling towers and wet-recirculating technology are used in geothermal power plants. Water requirements for geothermal plants can range from 1700 to 4000 gallons per megawatt-hour, depending on the cooling technology employed. The majority of geothermal plants, however, have the option of using either geothermal fluid or freshwater for cooling; using geothermal fluids instead of freshwater obviously lessens the overall water impact of the plant.

6.3.2. Air Emissions

When it comes to air emissions, the difference between open- and closed-loop systems is crucial. Air emissions are negligible in closed-loop systems because the gases extracted from the well are injected back into the ground after losing heat, preventing them from being exposed to the atmosphere. Open-loop systems, on the other hand, release boron, ammonia, carbon dioxide, hydrogen sulfide, and methane.

The most frequent emission is hydrogen sulfide, which has a distinctly "rotten egg" smell. Sulfur dioxide (SO_2) is produced when hydrogen sulfide enters the atmosphere. This leads to the production of tiny acidic particulates, which can enter the bloodstream and result in lung and heart problems. In addition to harming crops, forests, and soils, sulfur dioxide also produces acid rain, which also corrodes lakes and streams. But compared to coal plants, which are the country's largest source of SO_2, geothermal plants emit about 30 times less SO_2 per megawatt-hour.

Mercury emissions from some geothermal plants must be reduced with the use of mercury filter technology. Although scrubbers can lower air emissions, they also create a sludge that is watery and contains various heavy metals such as arsenic, mercury, nickel, vanadium, sulfur, and silica compounds. Hazardous waste sites are frequently required for the disposal of this toxic sludge (Table 6.3).

Table 6.3. *Summary of the Various Categories of Direct Use Worldwide, Referred to the Period 1995–2010*

Summary of the Various Categories of Direct Use Worldwide, Referred to the Period 1995–2010.				
	2010	**2005**	**2000**	**1995**
Capacity (MWt)				
Geothermal Heat Pumps	33,134	15,384	5275	1854
Space Heating	5391	4366	3263	2579
Greenhouse Heating	1544	1404	1246	1085
Aquaculture Pond Heating	653	616	605	1097
Agricultural Drying	125	157	74	67
Industrial Uses	533	484	474	544
Bathing and Swimming	6700	5401	3957	1085
Cooling/Snow Melting	368	371	114	115
Others	42	86	137	238
Total	**48,493**	**28,269**	**15,145**	**8,664**
Utilization (TJ/yr)				
Geothermal Heat Pumps	200,149	87,503	23,275	14,617
Space Heating	63,025	55,256	42,926	38,230
Greenhouse Heating	23,264	20,661	17,864	15,742
Aquaculture Pond Heating	11,521	10,976	11,733	13,493
Agricultural Drying	1635	2013	1038	1124
Industrial Uses	11,745	10,868	10,220	10,120
Bathing and Swimming	109,410	83,018	79,546	15,742
Cooling/Snow Melting	2126	2032	1063	1124
Others	955	1045	3034	2249

Summary of the Various Categories of Direct Use Worldwide, Referred to the Period 1995–2010.				
	2010	**2005**	**2000**	**1995**
Total	**423,830**	**273,372**	**190,699**	**112,441**
Capacity Factor				
Geothermal Heat Pumps	0.19	0.18	0.14	0.25
Space Heating	0.37	0.40	0.42	0.47
Greenhouse Heating	0.48	0.47	0.45	0.46
Aquaculture Pond Heating	0.56	0.57	0.61	0.39
Agricultural Drying	0.41	0.41	0.44	0.53
Industrial Uses	0.70	0.71	0.68	0.59
Bathing and Swimming	0.52	0.49	0.64	0.46
Cooling/Snow Melting	0.18	0.18	0.30	0.31
Others	0.72	0.39	0.70	0.30
Total	**0.28**	**0.31**	**0.40**	**0.41**

Source: Lund JW. Direct Utilization of Geothermal Energy. Energies. 2010; 3(8):1443-1471. https://doi.org/10.3390/en3081443.

6.3.3. Land Use

A geothermal plant's land requirements vary according to the resource reservoir's characteristics, power capacity, energy conversion system type, cooling system type, well, and pipe system layout, substation requirements, and auxiliary building needs. Approximately 1517 megawatts is the capacity of the largest geothermal plant in the world, The Geysers, which occupies an area of about 78 square kilometers, or 13 acres per megawatt. Similar to geysers, a lot of geothermal sites are situated in isolated, delicate ecological regions; project developers need to consider this when designing their schemes.

Land subsidence, a phenomenon in which the land surface sinks, is sometimes caused by the removal of water from geothermal reservoirs. Most geothermal facilities address this risk by re-injecting wastewater back into geothermal reservoirs after the water's heat has been captured.

Geological "hot spots," which typically have higher levels of earthquake risk, are where hydrothermal plants are located. There is proof that the presence of hydrothermal plants can increase the frequency of earthquakes. Additionally, hot, dry rock with enhanced geothermal systems may raise the possibility of minor earthquakes. This technique uses a high-pressure water pump, akin to that of hydraulic fracturing for natural gas, to fracture underground hot rock reservoirs. The risk of earthquakes related to enhanced geothermal systems can be reduced by placing plants at a suitable distance from active fault lines. When a geothermal system is located close to a densely populated area, open communication and ongoing monitoring with the local community are also essential.

6.3.4. Life-Cycle Global Warming Emissions

Methane, a more potent greenhouse gas, makes up a smaller portion of the air emissions from open-loop geothermal systems, with carbon dioxide accounting for around 10% of total emissions. Open-loop systems are estimated to emit 0.1 pounds of carbon dioxide equivalent per kilowatt-hour, contributing to global warming. Although these gases are not released into the atmosphere in closed-loop systems, there are still some emissions related to the building of the plant and the surrounding infrastructure.

Enhanced geothermal systems have a life-cycle global warming emission of about 0.2 pounds of carbon dioxide equivalent per kilowatt-hour. These systems require energy to drill and pump water into hot rock reservoirs. To put this into perspective, life-cycle global warming emissions estimates for electricity generated by natural gas range from 0 to 2 pounds of carbon dioxide equivalent per kilowatt-hour, while estimates for electricity generated by coal range from 1 to 3 points of carbon dioxide equivalent per kilowatt-hour (Office of Energy Efficiency and Renewable Energy, 2013; U.S. Department of Energy, 2014).

6.4. CARBON FOOTPRINT REDUCTION

To mitigate climate change and make the transition to a sustainable future, reducing carbon footprints is essential. A person, organization, or activity's total greenhouse gas emissions, measured as their "carbon footprint," can be greatly reduced by implementing a number of different tactics. Using more renewable energy sources, like solar, wind, and hydro power, is one of the main strategies. In contrast to fossil fuels like coal and natural gas, these sources emit little to no greenhouse gases. Making the switch to renewable energy lowers dependency on dirty energy sources while also reducing emissions during the energy production process.

In addition to adopting renewable energy, improving energy efficiency is key to carbon footprint reduction. This can be achieved through upgrading to energy-efficient appliances, enhancing building insulation, and implementing smart grid technologies that optimize energy use. Such measures not only decrease energy consumption but also lower associated emissions.

Sustainable practices in transportation, such as using public transit, cycling, or electric vehicles, further contribute to lowering carbon footprints. Encouraging changes in personal behavior, like reducing air travel, conserving water, and minimizing waste through recycling and composting, also play significant roles.

Furthermore, adopting sustainable agricultural practices, such as reducing meat consumption and supporting local, organic farming, can reduce emissions related to food production and transportation.

On a broader scale, implementing carbon offset programs and investing in carbon capture technologies can help mitigate the remaining emissions that are difficult to eliminate.

Overall, a comprehensive approach involving renewable energy adoption, energy efficiency, sustainable practices, and technological advancements is essential for substantial carbon footprint reduction. By integrating these strategies, individuals, businesses, and governments can collectively work towards a more sustainable and climate-resilient future.

6.4.1. Analysis of Geothermal Energy's Emissions Savings

As a major contributor to the reduction of greenhouse gas emissions and the advancement of sustainable energy solutions, geothermal energy offers significant emissions savings when compared to conventional fossil fuel sources. To produce electricity and provide direct heating, geothermal power uses heat that is found beneath the surface of the Earth. When operating, geothermal energy systems emit very little direct carbon dioxide, in contrast to fossil fuels.

The main reason geothermal energy emits fewer emissions is because of its small carbon footprint during operation. Only 10 to 20 grams of CO_2-equivalent per kilowatt-hour (g CO_2e/kWh) of electricity produced is the typical emissions from geothermal power plants. Compared to natural gas plants, which generate roughly 450 to 550 g CO_2e/kWh, and coal-fired power plants, which release between 800 and 1,000 g CO_2e/kWh, this is much lower.

The reason for these decreases is that geothermal energy doesn't burn fossil fuels; instead, it uses heat that is stored in the Earth's crust. When compared to the emissions from fossil fuel-based power generation, the construction and maintenance of power plants is the primary source of emissions in the geothermal energy industry. Furthermore, by utilizing geothermal resources effectively and employing management techniques that reduce land use and environmental disturbance, geothermal systems can have a reduced environmental impact.

Geothermal energy also provides a stable and continuous power supply, contributing to a reliable low-carbon energy mix. By displacing fossil fuel-based power generation, geothermal energy reduces overall greenhouse gas emissions and supports efforts to meet climate goals.

Overall, geothermal energy stands out for its low emissions and its ability to provide consistent, renewable energy, making it an essential component of a sustainable energy strategy aimed at reducing global carbon footprints and addressing climate change.

6.4.2. Challenges: Resource Location, Sustainability, and Potential Environmental Impacts.

Geothermal energy, despite its substantial benefits for reducing greenhouse gas emissions, faces several challenges related to resource location, sustainability, and potential environmental impact

1. **Resource Location**: Geothermal energy resources are geographically specific, often found in regions with significant tectonic activity, such as volcanic areas or tectonic plate boundaries. This limitation means that geothermal power plants are typically concentrated in certain areas, such as Iceland, parts of the United States (e.g., California, and Nevada), and parts of East Africa. This uneven distribution can pose challenges for expanding geothermal energy to regions that lack accessible resources, potentially limiting its widespread adoption.
2. **Sustainability**: While geothermal energy is considered renewable, its sustainability depends on the management of geothermal reservoirs. Over-extraction of geothermal fluids can lead to resource depletion, reduced heat output, and subsidence. To ensure long-term sustainability, it is crucial to implement effective reservoir management practices, such as monitoring fluid levels, reinjecting spent fluids, and carefully controlling extraction rates. Proper management can help maintain the geothermal resource's health and extend its operational lifespan.
3. **Potential Environmental Impacts**: Although geothermal energy has a relatively low environmental footprint, it is not without impacts. Drilling and extraction processes can result in land disturbance and the release of trace gases, such as hydrogen sulfide and carbon dioxide. Additionally, large-scale geothermal projects can lead to

local environmental issues, such as habitat disruption or changes in groundwater levels. The development of geothermal plants must consider these potential impacts and incorporate measures to mitigate them, such as conducting thorough environmental impact assessments and implementing best practices for site management.

Addressing these challenges requires a balanced approach that combines technological innovation, rigorous environmental safeguards, and effective resource management to maximize the benefits of geothermal energy while minimizing its potential drawbacks (Office of Energy Efficiency and Renewable Energy, 2013; U.S. Department of Energy, 2014).

REFERENCES

1. Baeumont, A. (n.d.). *Geothermal Energy Plant* [Digital image]. Flickr. Retrieved from: https://www.flickr.com (accessed on 9 November 2024).
2. Beardsmore, G. R., & Cull, J. P. (2001). *Crustal Heat Flow: A Guide to Measurement and Modelling*. Cambridge University Press.
3. Burley, A. J., Edmunds, W. M., & Gale, I. N. (1984). *Investigation of the Geothermal Potential of the UK: Catalogue of Geothermal Data for the Land Area of the United Kingdom*.
4. Chowdhury, P. (2009). *Old Faithful Erupts* [Digital image]. Flickr. Retrieved from: https://www.flickr.com (accessed on 9 November 2024).
5. Gupta, H. K., & Roy, S. (2007). *Geothermal Energy: An Alternative Resource for the 21st Century*. Elsevier.
6. Lund, J. W., & Boyd, T. L. (2016). Direct utilization of geothermal energy 2015 worldwide review. *Geothermics, 60*, 66–93. https://doi.org/10.1016/j.geothermics.2015.09.008.
7. Renewable Energy World. (2014). Geothermal electricity production. *Geothermal Energy*. Retrieved from: https://www.renewableenergyworld.com (accessed on 9 November 2024).
8. U.S. Department of Energy. (n.d.). Geothermal–Energy explained, your guide to understanding energy. *Energy Information Administration*. Retrieved from: https://www.eia.gov/renewable/ (accessed on 9 November 2024).
9. U.S. Department of Energy. (n.d.). Geothermal. Office of Energy Efficiency & Renewable Energy. Retrieved from: https://www.energy.gov/eere/geothermal (accessed on 9 November 2024).

Chapter 7

Ocean Energy

Contents

INTRODUCTION

Ocean energy offers a promising and varied range of renewable energy sources since it is derived from the vast and dynamic forces of the sea. It includes a number of technologies intended to harness the energy of thermal gradients, waves, tides, and ocean currents. Among them is tidal energy, which uses turbines installed in tidal streams or estuaries to harness the rise and fall of the tides to produce electricity. While ocean current energy uses the constant flow of ocean currents to generate power continuously, wave energy uses the movement of surface waves to power mechanical systems or generators. Furthermore, through thermodynamic cycles, ocean thermal energy uses the temperature differential between warmer surface waters and colder deep waters to produce electricity. The consistency and predictability of ocean energy is one of its main benefits. In contrast to more erratic renewable energy sources like wind and solar power, tides, and ocean currents offer a stable energy source. This predictability, especially in coastal areas, can help create a more dependable and consistent energy supply.

Ocean energy does, however, also confront formidable obstacles. Due to the challenging marine environment, which can hasten equipment wear and tear, high installation and maintenance costs are significant obstacles. Environmental effects need to be carefully managed through comprehensive impact assessments and monitoring, including effects on local wildlife and marine ecosystems. Ocean energy has a lot of promise for producing clean, sustainable power despite these obstacles. Technology has the potential to be extremely important in diversifying the energy mix and lowering reliance on fossil fuels as costs come down and advancements occur. To maximize the effectiveness, lower expenses, and lessen the environmental impact of ocean energy technologies, ongoing research and development are imperative.

7.1. BASIC CONCEPT OF OCEAN ENERGY

More energy is needed in the world today than it has ever needed. There is an increasing demand for energy due to the expanding industries and population. They are now a necessary part of our everyday existence. But we also have to be mindful of the fact that this consumption shouldn't have a significant impact on the planet's ecological balance. There are studies and reports that focus on the consequences of using fossil fuels. The common man's sense appears to have been impacted, and they are now utilizing unconventional energy sources. Over 70% of the earth's surface is covered in water. It looks blue from the space because of this. Because there are seven major oceans on the planet, it is also known as the "blue planet." Oceans have shown to be a promising energy source. It is an energy source that is renewable. When compared to other unconventional energy sources like solar energy, it is still less common. Our scientists are engaged in the advancement and utilization of ocean energy. When compared to other global scales, the amount of ocean energy produced nowadays is minuscule.

7.1.1. What Is Ocean Energy?

Any renewable energy source that comes from the sea is referred to as ocean energy. Ocean technology comes in three primary flavors: wave, tidal, and ocean thermal. Every type of ocean energy is still in its infancy as a commercial product. The cost of wave energy is still higher than that of other ocean technologies. While tidal stream has been demonstrated at a pilot scale, tidal range (explained below) has been implemented in places throughout the world where there is a significant tidal resource.

7.1.2. How Does It Work?

Wave energy is generated by converting the energy within ocean waves (swells) into electricity. There are many different wave energy technologies being developed and studied to convert wave energy into electricity.

Tidal energy comes in two forms, both of which generate electricity:

1. Tidal range technologies harvest the potential energy created by the height difference between high and low tides. Barrages (dams) harvest tidal energy from different ranges.
2. Tidal stream (or current) technologies capture the kinetic energy of currents flowing in and out of tidal areas (such as seashores). Tidal stream devices operate in arrays, similar to wind turbines.

Ocean thermal energy is generated by converting the temperature difference between the ocean's surface water and deeper water into energy. Ocean thermal energy conversion (OTEC) plants may be land-based as well as floating or grazing.

7.2. TIDAL ENERGY

The natural rise and fall of ocean tides and currents provides the energy source for tidal energy, which is renewable. Among these technologies are paddles and turbines. Tidal energy is another form of ocean energy that can be trapped in reservoirs behind dams when tides come inshore. In actuality, tidal power is a type of hydropower that takes advantage of the water movement brought on by tidal currents or the rise and fall of sea levels. In locations with strong tidal movements, tidal energy generators—large underwater turbines intended to harness the kinetic motion of the ebbing and surging of ocean tides to generate electricity—are used to generate tidal energy. The vast expanse of the oceans presents a huge potential for tidal power in the production of electricity in the future. The long-standing recognition of tidal power's potential dates back to the 11th century, when small dams were constructed along oceanic regions. Tidal power projects are more costly than river dams, though, because they require the construction of large structures in a challenging saltwater environment. Despite its enormous potential, tidal power hasn't yet become one of the most popular renewable energy sources due primarily to cost effectiveness. There aren't many places on Earth that can meet the requirements for tidal power because it requires extremely large tide increases—at least 16 feet between low and high tide—to function at a sufficient level. This tidal power plant has a capacity that is about one-fifth that of a conventional nuclear or coal-powered plant. The primary issue with all tidal power plants is that they can only produce energy for 10 hours a day, during which the tide is either in or out. The fact that tides are completely predictable, on the other hand, allows us to schedule the operation of other power plants during the periods when the tidal station is not in operation—something that is not possible with certain other renewable energy sources, such as wind energy (Figure 7.1).

Figure 7.1. *Sihwa lake tidal power station in Gyeonggi province, South Korea.*

Source: By핑크로즈otherversions=–http://blog.naver.com/pfmusic/10122410431, CC BY 2.0 kr, https://commons.wikimedia.org/w/index.php?curid=41208311.

Although tidal energy has many benefits—it is a renewable energy source because tides will always ebb and flow, it emits no greenhouse gases or waste, it doesn't require fuel to operate, and because tides are completely predictable, it can produce electricity reliably and is inexpensive to maintain once installed—it also has some drawbacks. Cost effectiveness is still a major concern because constructing one of these power plants takes up a lot of space and causes environmental issues because it drastically alters the local ecosystem and has an impact on many species, particularly birds that depend on the tide to expose mud flats where they can find food. In addition, as was previously mentioned, there is the issue of the short working day—roughly 10 hours—when the tide is truly out (Chiles, 2009; Lewis, 2011).

7.2.1. Tidal Energy Generation

We can use the potential energy of the water contained in the daily movement of the rising and falling sea levels to generate electricity because the position of the earth and the moon with respect to the sun changes throughout the year. There are numerous similarities between the hydroelectric generation we examined in the hydro energy and the electricity generated by tides. This time, however, there is a difference: rather than only flowing forward, the water enters and exits the turbines in both directions (Figure 7.2).

Figure 7.2. *Tidal turbines create energy efficiently and with almost no environmental impact.*

Source: By Tswgb – Own work, Public Domain, https://commons.wikimedia.org/w/index.php?curid=3182853.

Similar to hydro energy, tidal energy converts moving water into clean energy. Large volumes of kinetic energy are present in the motion of the tidal water, which is propelled by gravity and forms powerful tidal currents known as tidal streams. Water flowing down a river or stream is not much different from the daily ebbing and flowing of the ocean tides along a coastline and into and out of small inlets, bays, or coastal basins.

The process of harnessing the movement of seawater using waterwheels and turbines is comparable to that of producing hydroelectricity. However, in a tidal energy system, the sea water can flow both ways, allowing power to be produced both when the water is coming in and going out. As a result, tidal generators are made to generate electricity regardless of which way the rotor blades spin. Reversible electrical generators are, nevertheless, more costly than single-direction generators.

7.2.2. Types of Tidal Energy Systems

The power of tides can be used in many different ways. Using a "barrage," or a particular kind of dam, to obstruct receding water during ebb periods, tidal barrage power systems take advantage of variations in high and low tides. When the tide is low, the water behind the barrage is released and flows through a turbine to produce energy. Ocean currents can be used by tidal stream power systems to power turbines, especially in locations near islands or coasts where the currents are swift. They can be set up as tidal turbines, which resemble underwater wind turbines (see wind power), or as tidal fences, where turbines are stretched across a channel (Figure 7.3).

Figure 7.3. *The world's first commercial-scale and grid-connected tidal stream generator – SeaGen – in Strangford Lough.*

Source: By Fundy (Fundy) – Own work, CC BY-SA 3.0, https://commons.wikimedia.org/w/index.php?curid=4074928.

7.2.2.1. Advantages of Tidal Energy

1. Tidal energy is a renewable energy resource because the energy it produces is free and clean as no fuel is needed and no waste bi-products are produced.
2. Tidal energy has the potential to produce a great deal of free and green energy.
3. Tidal energy is not expensive to operate and maintain compared to other forms of renewable energy.
4. Low visual impact as the tidal turbines are mainly if not totally submerged beneath the water.
5. Low noise pollution as any sound generated is transmitted through the water.
6. High predictability as high and low tides can be predicted years in advance, unlike wind.
7. Tidal barrages protect against flooding and land damage.

Large tidal reservoirs have multiple uses and can create recreational lakes and areas where before there were none (Figure 7.4).

Figure 7.4. *Is tidal energy.*

Source: By Kches16414 – Own work, CC BY-SA 4.0, https://commons.wikimedia.org/w/index.php?curid=95341987.

7.2.2.2. Disadvantages of Tidal Energy

1. Tidal energy is not always a constant energy source as it depends on the strength and flow of the tides which themselves are affected by the gravitational effects of the moon and the sun.
2. Tidal Energy requires a suitable site, where the tides and tidal streams are consistently strong.
3. Must be able to withstand forces of nature resulting in high capital, construction, and maintenance costs.
4. High power distribution costs to send the generated power from the submerged devices to the land using long underwater cables.
5. Intermittent power generation, only generates power 10 hours a day during the ebb and flow of the tides.
6. Changes to the estuary ecosystem and an increase in coastal erosion where the tides are concentrated.
7. Buildup of silt, sediments, and pollutants within the tidal barrage from rivers and streams flowing into basin as it is unable to flow out into the sea.
8. Danger to fish and other sea life as they get stuck in the barrage or sucked through the tidal turbine blades (Chiles, 2009; Lewis, 2011).

7.3. WAVE ENERGY

Wave Energy also known as Ocean Wave Energy, is another type of ocean based renewable energy source that uses the power of the waves to generate electricity. The topic of renewable energy is always relevant, particularly in a world where fossil fuels predominate. In the modern world, renewable energy is a hot topic because it is sustainable, limitless, and doesn't release greenhouse gases that are bad for the environment or people's health. Wave energy is a well-known example of renewable energy.

Energy extracted from ocean or sea waves is referred to as wave energy, ocean energy, or sea wave energy. Wave energy technologies harness the vast amounts of kinetic energy found in the intense vertical motion of surface ocean waves to perform practical tasks like desalinating water, pumping water into reservoirs, and producing electricity (Table 7.1).

Table 7.1. *General Info of the Studied Wave and Tidal Systems Studied by Life Cycle Assessments (LCAs)*

Reference	Device	Region	Installed capacity	Capacity factor	ISO Smartcards	
					14040	14044
Cavallaro et al. (2007)	Kobold	Italy	160 kW	N.A.	-	-
Douglas et al. (2008)	Seagen	UK	1.2 MW	48%	√	-
Rule et al. (2009)	Kaipara Harbor	New Zealand	200 MW	37%	-	√
Howell et al. (2013)	DeepGen; OpenHydro; Scot Renewables SR2000; Flumill	UK	1 MW; 2 MW; 2 MW; 2 MW	N.A.	√	√
Hans et al. (2007)	Wave Dragon	Denmark	7 MW	N.A.	-	-
Parker et al. (2007)	Pelamis	UK	750 kW	N.A.	√	-
Thomson et al. (2011a)	Pelamis	UK	750 kW	N.A.	√	-
Thomson et al. (2011b)	Pelamis	UK	750 kW	N.A.	√	-
Thomson et al. (2019)	Pelamis	UK	750 kW	45%	-	√
Dahlsten et al. (2009)	Seabased	Hypothetical	20 MW	N.A.	√	√
Walker et al. (2011)	Oyster	UK	315 kW	N.A.	-	-
Ombach et al. (2014)	Wave Star	Denmark	1000 kW	N.A.	14000	-
Zhai et al. (2018)	BRD	China	10 kW	50%	√	-

Reference	**Device**	**Region**	**Installed capacity**	**Capacity factor**	**ISO Smartcards**	
					14040	**14044**
Elginoz et al. (2017)	MUP farm	Hypotheti-cal	265.5 MW	Various	√	-
Curto et al. (2018)	DEIM	Italy	30 kW	N.A.	√	√
Patrizi et al. (2019)	OBREC	Italy	3 kW	N.A.	-	-
Uihlein et al. (2016)	Various	Unspecified	500kW-1 MW	34%; 0%	-	-
Douziech et al. (2016)	Annapolis; SeaGen; HS1000; HydraTidal; Oyster800	Canada; Ireland; UK; Norway; UK	20 MW; 1200 kW; 1000 kW; 1500 kW; 800 kW	N.A.	-	-

Source: Zhang X, Zhang L, Yuan Y, Zhai Q. Life Cycle Assessment on Wave and Tidal Energy Systems: A Review of Current Methodological Practice. International Journal of Environmental Research and Public Health. 2020; 17(5):1604. https://doi.org/10.3390/ijerph17051604.

In essence, wave energy or wave power is power extracted from waves. Waves receive energy from the wind when it blows across the sea surface. They are strong energy sources. Wave height, wave speed, wavelength, and water density are used to calculate the energy output. It can generate more power when the waves are stronger. The energy that has been captured can then be used to pump water, power plants, or generate electricity. The fact that wave generator plants are difficult to use as a source of electricity accounts for their small number worldwide.

The wave's kinetic energy powers a turbine that is connected to a generator to generate electricity. But because open waters can be stormy and violent, the very energy that the wave energy machines are meant to capture ends up destroying them.

The most basic definition of an ocean wave is the vertical movement of seawater that varies sinusoidally over time. There are low points known as troughs and high points known as crests in this sinusoidal wave. A wave's amplitude, or height, is measured from peak to peak, or the difference in height between the crest and trough. The center of these two points represents the wave's actual height when there is no movement in the water or a calm sea.

The amplitude of an ocean wave depends on the weather conditions at that time, as the amplitude of a smooth wave, or swell, will be small in calm weather but much larger in stormy weather with strong gales as the sea water moves up and down.

The wave period (T), which is the interval between each subsequent crest or trough, is another crucial property in addition to the wave's amplitude. The interval of time in seconds between each wave crest is known as the wave period. Then, as each wave crashes onto the one in front of it, this time period may be very short in a stormy sea, but it may be very long in a gentle swell.

We can find the fundamental frequency of the ocean wave in relation to a static point by taking the reciprocal of this time (1/T). Harmonic waves are smaller periodic waves that are generated or superimposed upon this fundamental wave, such as reflected waves. Then, the fetch—the distance the wind blows over the open water—as well as the wind speed, duration, and water depth all affect the frequency and amplitude characteristics of a wind-generated wave.

Waves carry energy to a shoreline from their source, far out in the ocean, where storms create it. However, a typical ocean wave is more complex and irregular than a simple sinusoidal wave; it does not resemble a perfect sinusoid. Real sea waves are a mixture of waves with different frequencies, wave heights, and wave directions; the only thing that resembles a sinusoidal wave significantly more is the steady up-and-down movement of a heavy swell.

7.3.1. Wave Power Devices

Compared to ocean wind energy, ocean wave energy offers more available energy densities and is less variable and more predictable. Wave energy systems can be categorized as offshore, nearshore, or shoreline devices based on the distance between the energy conversion device and the shoreline. What distinguishes these three categories of energy extraction apparatuses from one another, then? Wave energy devices known as "shoreline devices" are those that are affixed to or integrated into the shoreline, meaning they can be in or out of the water.

The ability to extract wave power directly from the breaker zone and the waters immediately beyond it is what distinguishes nearshore devices, (i.e., in 20 meters of water). Utilizing the higher power wave profiles and higher energy densities found in deep water waves and surges, offshore or deep water devices are the furthest out to sea and extend beyond the breaker lines.

One of the advantages of offshore devices is that there is no need for significant coastal earthworks, as there is with onshore devices.

As most of the energy within a wave is contained near the surface and falls off sharply with depth. There is a surprising range of designs available that maximize the energy available for capture. These wave energy devices are either fixed bottom standing designs used in shallow water and which pierce the water surface, or fully floating devices that are used to capture the kinetic energy content of a wave's movement and convert each movement into electricity using a generator.

There are currently four basic "capture" methods:

1. **Point Absorbers** – these are small vertical devices either fixed directly to the ocean floor or tethered via a chain that absorb the waves energy from all directions. These devices generate electricity from the bobbing or pitching action of a floating device. Typical wave energy devices include, floating buoys, floating bags, ducks, and articulated rafts, so forth. These devices convert the up-and-down pitching motion of the waves into rotary movements, or oscillatory movements in a variety of devices to generate electricity. One of the advantages of floating devices over fixed devices it that they can be deployed in deeper water, where the wave energy is greater.
2. **Wave Attenuators** – also known as "linear absorbers," are long horizontal semi-submerged snake-like devices that are oriented parallel to the direction of the waves. A wave attenuator is composed of a series of cylindrical sections linked together by flexible hinged joints that allow these individual sections to rotate and yaw relative to each other. The wave-induced motion of the device is used to pressurize a hydraulic piston, called a ram, which forces high pressure oil through smoothing accumulators to turn a hydraulic turbine generator producing electricity. Then wave attenuators convert the oscillating movement of a wave into hydraulic pressure.
3. **Oscillating Water Column** – is a partly submerged chamber fixed directly at the shoreline which converts wave energy into air pressure. The structure could be a natural cave with a blow hole or a man-made chamber or duct with a wind turbine generator located at the top well above the water surface. The structure is built perpendicular to the waves so that the ebbing and flowing motion of the waves force the trapped water inside the chamber to oscillate in the vertical direction.

As the waves enter and exit the chamber, the water column moves up and down and acts like a piston on the air above the surface of the water, pushing it back and forth. This air is compressed and decompressed by this movement and is channeled through a wind turbine generator to produce electricity. The speed of air in the duct can be enhanced by making the cross-sectional area of the duct much less than that of the column.

4. **Overtopping Devices** – also known as "spill-over" devices, are either fixed or floating structures that use ramps and tapered sides positioned perpendicular to the waves. The sea waves are driven up the ramp and over the sides filling-up a small tidal reservoir which is located 2 to 3 meters above sea level. The potential energy of the water trapped inside the reservoir is then extracted by returning the water back to the sea through a low head Kaplan turbine generator to produce electricity.

Then overtopping devices convert the potential energy available in the head of water into mechanical energy. The disadvantage of onshore overtopping schemes is that they have a relatively low power output and are only suitable for sites where there is a deep water shoreline and a low tidal range of less than about a meter (Table 7.2).

Table 7.2. *Electricity Statistics for EU-MED Countries (in GWh)*

EU-MED Country	Year	Total Net Production	Imports	Exports	Energy Supplied
Cyprus	2014	4145	-	-	4145
	2015	4312	-	-	4312
	2016	4634	-	-	4634
France	2014	540,285	7873	75,063	465,138
	2015	544,693	9961	74,024	473,754
	2016	531,147	20,790	61,406	483,798
Greece	2014	46,702	9461	642	55,334
	2015	47,806	11,081	1473	57,340
	2016	48,052	9833	1037	56,815
Italy	2014	269,148	46,747	3031	310,535
	2015	272,428	50,849	4471	316,897
	2016	275,649	43,181	6155	310,251

EU-MED Country	Year	Total Net Production	Imports	Exports	Energy Supplied
Malta	2014	2137	-	-	2137
	2015	1239	-	-	2293
	2016	848	-	-	2375
Spain	2014	268,380	12,310	15,716	259,772
	2015	269,750	14,956	15,089	265,097
	2016	264,356	21,845	14,178	267,204
Slovenia	2014	16,486	7254	9997	13,380
	2015	14,187	9045	9093	13,759
	2016	15,432	8358	9536	13,881

Source: Eurostat (online data code: nrg_105a, nrg_105m).

Source: Soukissian TH, Denaxa D, Karathanasi F, Prospathopoulos A, Sarantakos K, Iona A, Georgantas K, Mavrakos S. Marine Renewable Energy in the Mediterranean Sea: Status and Perspectives. Energies. 2017; 10(10):1512. https://doi.org/10.3390/en10101512.

The idea of harnessing the tremendous power of the ocean waves is not new. Like other forms of hydro power, wave energy does not require the burning of fossil fuels, which can pollute the air, contributing to acid rain and global warming. The energy is entirely clean and endlessly renewable. Wave power has many advantages compared to other forms of renewable energy with its main advantage being that it is predictable.

However, like many other forms of renewable energy, ocean wave energy also has its disadvantages such as its inflexible generation time dependent upon the tides, the visual impact of wave devices on the sea surface, as well as the threat of collision with shipping and navigation (Chiles, 2009; Lewis, 2011).

7.3.2. Advantages of Wave Energy

7.3.2.1. It's Highly Predictable

The wave arrival pattern is highly predictable. They arrive day and night and harbor more energy than other renewable sources like wind and solar. Wind energy and solar energy, on the other hand, are highly unpredictable. Wind speeds die down unexpectedly, which affects the generation of electricity. Solar energy depends upon exposure to the sun, which means cloud coverage and night hours significantly reduce this exposure leading to less efficiency.

7.3.2.2. It's a Renewable form of Energy

Renewable means it's an endless resource. It does not need man's intervention to continue existing. No one has dared to suggest that the oceans and seas will disappear someday. Humans will continue harnessing it to the very end. This aspect makes wave energy a reliable and efficient energy resource.

7.3.2.3. Wave Energy Is Eco-Friendly

Wave energy is a completely clean energy source, which means, it does not emit dangerous greenhouse gasses to the atmosphere. Fossil fuels, for instance, oil, coal, and natural gas contribute mightily to environmental pollution because they release dangerous greenhouse gasses including carbon dioxide, nitrous oxide, methane, and ozone into the atmosphere.

7.3.2.4. Creation of Green Jobs

Communities living in remote areas and declining industries like the ship-building industry bear the biggest brunt of unemployment and economic unsustainability due to lack of electricity. The wave energy sector has the potential to create numerous green opportunities for remote and urban populations alike because remote areas that are not able to be reached by conventional electricity supply are well catered by wave power.

7.3.2.5. Exponential Growth of Remote Areas

The wave energy harnessed can to channeled to remote locations, and this means springing up of industries and businesses. These remote areas will witness strong economic growth moving forward.

7.3.2.6. Security of Energy Supply

Setting up a strong wave energy infrastructure can enormously help a country from over-dependence on fossil fuels. The fossil fuel market is largely volatile and could hurt a country's economy if shortage occurs. Wave energy is the surefire way to bridge this volatility gap since it's cheap, reliable, and efficient.

7.3.2.7. Land Remains Undamaged

Wave energy plants can be situated offshore alleviating any risk that comes along with these plants situated onshore like soil pollution. Also, the land remains in its natural state unlike fossil fuel extraction, which requires high levels of excavation that leaves land heavily damaged.

7.3.3. Disadvantages of Wave Energy

7.3.3.1. High upfront Capital Costs

Wave energy plant construction requires a significant financial investment. Some factors that drive up the cost of wave energy are energy plant maintenance, grid connection, wave resources, anticipated reduction in energy costs once infrastructure is operational, and technology shelf life. Given that wave energy is still in its infancy, estimating the true cost is also challenging.

7.3.3.2. Variability in Wave Magnitude Can Damage Equipment

The wave magnitude is so unpredictable in the seas. Sometimes it comes with a vengeance and could cause heavy wear and tear to the wave energy generation turbines. Damage to this equipment can be costly in terms of repair. It would also mean stalling of electricity supply.

7.3.3.3. Damage to Sea Life Ecosystem

Wave energy projects offshore are far more advanced than those onshore. Dredging, interconnections, platforms, cables, turbines, and much more are all part of the projects. Shallow waters are ideal for most marine life's breeding and resting grounds from an ecological perspective. Hence, the marine ecosystem is significantly impacted by the development and maintenance of wave energy plants. Unintentional spills or leaks from the hydraulic fluids in the plants could contaminate the water and kill marine life (Chiles, 2009; Lewis, 2011).

7.4. OCEAN CURRENT ENERGY

A significant amount of water and energy are transported across the earth's oceans by the comparatively steady flow of ocean currents. Technologies are being developed so that ocean currents may provide a source of clean, renewable energy that can be extracted and transformed into usable power, even though there hasn't been any commercial development in the US.

Ocean currents have intricate patterns and routes that are influenced by many factors, including wind, temperature, ocean floor topography, earth's rotation, and water salinity. While some ocean currents are caused by variations in the water column's salinity and density, the majority of ocean currents are propelled by surface waters heated by the sun and wind.

The global ocean circulation, which is influenced by density gradients produced by surface heat and freshwater fluxes, includes the thermohaline circulation. The terms "thermo-," which refers to temperature, and "-haline," which refers to salt content, combine to define the term thermohaline, which describes the density of ocean water. From the equatorial Atlantic Ocean, wind-driven surface currents like the Gulf Stream move poleward, cooling as they go and finally sinking at high latitudes to form the North Atlantic Deep Water. The ocean basins are then filled with this dense water. The oldest waters in the world upwell in the North Pacific, but the majority of it does so in the Southern Ocean. As a result, water diffuses and mixes between the ocean basins, lessening their disparities and transforming the oceans into a single, global system.

The water masses carry energy in the form of heat as they travel. Unlike tidal currents along the shore, ocean currents flow in a relatively constant and directed manner. Because of the density of water, ocean currents move more slowly than wind speeds, but they still carry a lot of energy. Water moves at 12 miles per hour with the same amount of force as a constant wind of 110 mph because it is more than 800 times denser than air. Ocean currents have a tremendous amount of energy that can be captured and transformed into a useful form because of this physical characteristic.

7.4.1. Ocean Current Energy Technologies

While marine current energy is still in its infancy, the United States and other nations are pursuing ocean current energy. Ocean current power has the most untapped energy potential compared to wind, wave, and tidal resources, but its technology is also the least developed. Just a handful of prototypes and demonstration units have undergone testing, and there are currently no grid-connected turbines in operation on a commercial scale. In near-shore environments, more sophisticated technologies have been developed for use with tidal currents. Many distinct technology concepts are currently being developed. Similar to wind turbines, prototype horizontal axis turbines have been constructed and tested; over the next five to seven years, commercial development of these devices is most likely to occur. While there is still much to learn about ocean current technology, a number of tidal and in-stream current turbine applications are almost ready for commercialization. These gadgets make use of the regular water flow from freshwater rivers or the daily tidal cycles in near-shore ocean environments (Table 7.3).

Table 7.3. *Comparison of Various Renewable Energy Sources*

Renew-able	**Predict-ability**	**Visual**	**Environmental**	**Capital**
Source		**Impact**	**Impact**	**Cost**
Wind	No	High	Medium	High
Solar	No	High	Low	High
Hydro	Yes	High	Medium	High
Wave	No	Medium	Low	High
Tidal range	Yes	High	Medium	High
Tidal current	Yes	Low	Low	High

Source: Rashid H, Benbouzid M, Titah-Benbouzid H, Amirat Y, Mamoune A. Tidal Stream Turbine Biofouling Detection and Estimation: A Review-Based Roadmap. Journal of Marine Science and Engineering. 2023; 11(5):908. https://doi.org/10.3390/jmse11050908.

The greatest places to capture the energy of ocean currents are in the spaces between islands, along heavily tidal-influenced coasts, or in areas where the temperature differential between the warm surface water and the cold deep water is 20□ or greater. The most widely recommended technology is marine turbines. To generate three times as much power, the blades of these turbines only need to be about 20 meters long, or one-third the size of a wind generator. As with offshore wind towers, each turbine will be installed atop a tower that will be connected to the grid by an underwater cable. To alert ships, the towers will protrude above the water's surface and be illuminated. Additionally, these towers are designed to be lifted out of the water for maintenance purposes and to remove growths such as seaweed from the blades.

The kinetic energy of moving water is converted into electrical power by current energy technologies, also known as tidal or hydrokinetic technologies. Modern energy technologies use the ocean's horizontal current flow to drive a generator, which transforms mechanical energy into electrical power. Current energy devices are frequently wind turbine-like rotating machines with a rotor that rotates in response to water currents, with the rotational speed being proportionate to the water's velocity. The rotor can be enclosed in a duct that directs the flow of water and current, or it can have an open design similar to a wind turbine. There are four primary categories of current energy converters.

7.4.1.1. Horizontal Axis Turbines

Horizontal axis turbines typically look similar to wind turbines. They harvest kinetic energy from the moving water in the same way that wind turbines extract energy from the moving air.

7.4.1.2. Ducted Horizontal Axis Turbines

The horizontal rotor is enclosed inside a duct. These ducts are often funnel-shaped and work by concentrating the current flow through the turbine. This technology may allow for function over a wider range of current velocities, thus producing more electricity per unit of rotor area.

7.4.1.3. Vertical Axis Turbines

In vertical axis turbines, the axis of the rotor is situated perpendicular to the current flow. These turbines may also be enclosed within a duct.

7.4.1.4. Oscillating Hydrofoils

Oscillating hydrofoils pivot in response to tidal currents flowing over a wing or flap-like structure. The special situating achieve with this design drive fluid in a hydraulic system to generate electricity.

7.4.2. Challenges

For ocean current energy to be used successfully at a commercial scale, a number of engineering and technical challenges need to be addressed, including:

1. Avoidance of cavitations (bubble formation).
2. Prevention of marine growth buildup.
3. High maintenance costs.
4. Corrosion resistance.
5. Generation of electromagnetic fields (EMF).
6. Toxicity of paints, lubricants, and antifouling coatings.
7. Interference with animal movements and migrations, including entanglement and strike by rotor blades or other moving parts.

To calculate the number of joules available for extraction by turbines in all the oceans of the entire world, lets speculate that the ocean ubiquitously is moving at a speed in a certain direction, equal to that of the Gulf Stream; about 2 meters per second. Therefore, each cubic meter of water has a kinetic energy 1/2 Mv^2 = 2000 joules. This may be an extreme overestimate since most of the ocean moves much more slowly than the Gulf Stream, but assuming the larger number, we then multiply it by the number of cubic meters of all the oceans of the entire world: The radius of the earth is R = 6371 km, the average depth of the ocean is d = 4.27 km, and assuming that 70% of the world's surface is ocean, the total volume of water in the world's oceans is:

$$\Omega = 0.7 \times 4\pi R^2 d$$

$$= 0.7 \times 4 \pi \times (6.37 \times 10^6 \text{m})^2 \times 4.27 \times 10^3 \text{ m}$$

$$= 1.52 \times 10^{18} \text{ m}^3$$

Therefore, the total amount of energy available for extraction in the world's ocean is 2000 joules/m^3 x 1.52×10^{18} m^3 = 3.04×10^{21} joules. When you compare this total to the annual energy demand per year for the entire world (Chiles, 2009; Lewis, 2011).

7.5. OCEAN THERMAL ENERGY CONVERSION

The process known as Ocean Thermal Energy Conversion (OTEC) uses the temperature differential between warm tropical surface waters and deep, cold ocean water to generate electricity. OTEC plants use a lot of surface and deep-cold seawater to run a power cycle and generate electricity. OTEC is a clean energy source that is reliable around-the-clock, environmentally friendly, and able to produce large amounts of energy.

Ocean thermal energy conversion, or OTEC for short, is a renewable energy source that creates a heat engine by taking advantage of the temperature differences in the ocean at different depths. This heat energy is transformed into a mechanical output by the heat engine. Ocean thermal energy conversion systems come in three different varieties: closed, open, and hybrid.

OTEC systems come in three varieties at the moment: closed, open, and hybrid. A liquid with a very low boiling point powers a turbine in a closed cycle system to produce energy. Warm ocean surface water is sent through a heat exchanger to evaporate to achieve this. The generator moves as the vapor expands. Using a different heat exchanger, cold ocean water that has been drawn from a much deeper depth is reduced from vapor to liquid. Since the water from the closest ocean to the equator is usually the warmest, an open cycle system uses that water.

A low pressure container filled with warm ocean water will start to boil. The salt stays behind as the steam expands and escapes the low pressure container. A low pressure engine that is connected to an apparatus that generates electricity is powered by the steam. By subjecting it to the chilly temperatures produced by ocean water at a lower depth, it is transformed back into liquid form.

The elements of the first two approaches are combined in the hybrid system. Steam is produced by drawing warm ocean water into a container. A low boiling point liquid is subsequently turned into vapor by the steam, which drives a turbine to produce energy (Figure 7.5).

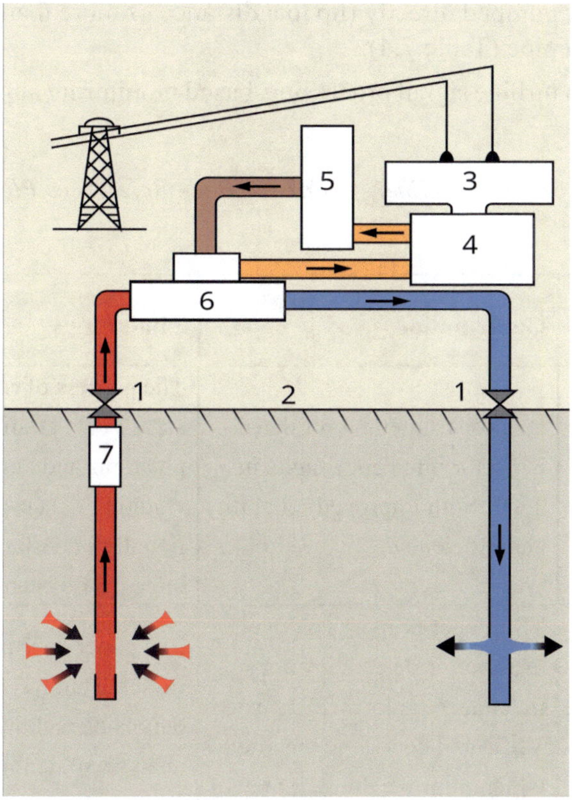

Figure 7.5. *Electricity generation in a vapor-dominated hydrothermal system. Key: 1 – Wellheads, 2 – Ground surface, 3 – Generator, 4 – Turbine, 5 – Condenser, 6 – Heat exchanger, 7 – Pump.*

Source: https://upload.wikimedia.org/wikipedia/commons/thumb/0/01/Geothermal_Binary_System.svg/1200px-Geothermal_Binary_System.svg.png.

7.5.1. Types of OTEC Technology

Several options exist for OTEC technology depending of the location and the cycle used. Plants may be based on land, ocean shelf or floating facilities. The first two have the most advantages. Three kinds of cycles are used: the open cycle, the closed cycle and the hybrid cycle.

For every system, the fundamental idea is the same. A heat engine is positioned in between a reservoir with a high temperature and one with a low temperature. The engine transforms thermal energy into kinetic energy, just like a steam turbine does. To function, the cold sea water is either desalinated close to the sea bed or pumped directly (up to a distance of more than a kilometer) to the surface via a pipe (Table 7.4).

Tidal stream turbine signal processing-based monitoring approaches critical analysis.

Table 7.4. *Monitoring Tidal Energy (For the Back References Please Refer to the Cited Article)*

Proposed Approach	**Contributions**	**Limitations**
Angular resampling	Proposes a method for detecting rotor imbalance faults in TSTs, with improved reliability and efficiency.	The process of resampling the sensor data at different angular positions and analyzing the resulting data can be computationally intensive, especially for large TST systems.
Variational mode decomposition (VMD)	This paper proposes a novel method for detecting rotor imbalance faults in TSTs using VMD and denoising techniques which improve the accuracy of the fault detection method to improve the detection accuracy.	The paper does not provide an in-depth analysis of the used denoising techniques. Further analysis of denoising techniques could help identify the optimal denoising method.
Concordia transform	Proposing an advanced concordia transform for blade imbalance fault detection by using TST generator stator currents.	The proposed method lacks analysis of false positive rates, which may increase the operating costs of the TST system due to unnecessary maintenance and repairs.

Proposed Approach	**Contributions**	**Limitations**
Wavelet threshold denoising	This paper proposes a method to detect imbalance faults in TSTs under different flow velocity conditions.	The generalizability of the proposed method to different TST systems and operating conditions is not clear, and further evaluation is needed to understand the performance of the proposed method under different scenarios.
Continuous wavelet transform	Extracting frequency-domain features from the vibration signals generated by the blades of marine hydrokinetic turbines which can be used to accurately classify the blades as healthy or faulty.	Does not provide a detailed comparison of the proposed method with other existing methods for detecting blade faults in marine hydrokinetic turbines.
Bispectrum analysis	This paper presents experimental results that evaluate TSTs stator currents bispectrum analysis in detecting biofouling.	A detailed comparison of the proposed method with other existing methods for detecting biofouling in TSTs is not provided.
Higher-order spectra	This paper proposes detecting blade biofouling in a TST using higher-order spectra analysis of the stator currents of its permanent magnet synchronous generator.	The study focuses solely on the use of higher- order spectral analysis to detect biofouling of TST blades while other types of spectral analysis, such as wavelet analysis, can improve its performance.

Proposed Approach	**Contributions**	**Limitations**
Data normalization and empirical mode decomposition (EMD)	The proposed method normalizes TST generator stator current signals and then applies EMD to identify the presence and severity of imbalance faults including wave and turbulence conditions.	The proposed fault detection methods assume the model parameters are known. However, in practice, these parameters are generally unknown and should be estimated. The potential benefits of using Ensemble EMD (EEMD) to improve detection accuracy have not been investigated.
Integration methodology	The proposed method integrates the features of the TST generator stator voltage signal to detect imbalance faults. The detection process is based on two main steps: data conversion using Hilbert transform and extreme value searching, and then the imbalance fault signature extraction using frequency sequences subtraction.	A detailed comparison of the proposed method with other existing methods for detecting biofouling in TSTs is not provided. In addition, it is yet to be demonstrated whether voltage can be effectively used for detection, given that TSTs connected to the distribution grid will an imposed voltage

Source: Rashid H, Benbouzid M, Titah-Benbouzid H, Amirat Y, Mamoune A. Tidal Stream Turbine Biofouling Detection and Estimation: A Review-Based Roadmap. Journal of Marine Science and Engineering. 2023; 11(5):908. https://doi.org/10.3390/jmse11050908.

7.5.1.1. Closed-Cycle

This cycle uses a working fluid that goes through a full cycle of cooling and heating (due to its low boiling point). Warm water from the surface is injected into a heat exchanger, where the heat causes the working fluid to change state and evaporate. The vapor is directed toward the generator and turbine, which produce the electricity. The extremely cold water then causes the vapor to condense. After that, the liquid returns to the heat exchanger. India constructed a floating, closed-cycle plant with a 1 MW rated power.

7.5.1.2. Open-Cycle

Unlike the closed-cycle process, the open-cycle process uses sea water directly instead of passing through an intermediary fluid. It makes use of the tropical seas' warm surface. The evaporation process is carried out by placing the warm water in a low-pressure container. The steam is fed into the turbine that is connected to a generator, just like in previous cycles. The second effect is evaporation, which produces fresh water (desalinated water). The extremely cold water causes the steam to condense into liquid. Since the water is now safe to drink, it is captured, in contrast to the first cycle.

7.5.1.3. Hybrid Cycle

The last cycle proposes to combine both open and closed cycles. The warm water is vaporized in a vacuum chamber. The steam transforms working liquid into vapor (heat exchanger) which is directed in a turbine. As in the open-cycle the condensed water provides fresh water.

7.5.2. Applications of OTEC

The main goal of this system is to output electric power and, secondly, desalinated water (2-megawatt electric plant could produce about 4300 cubic meters of desalinated water each day) for some cycles. But OTEC technology offers others possibilities, like support for deep water mariculture (deep waters rich in nutrients) and air conditioning. It also can be used to produce ammonia (the working fluid), hydrogen, aluminum, methanol, and others chemicals.

7.5.2.1. Perspectives

A competitive market could be a target in the next decade for OTEC plants. The small island nations could benefit because the cost to generate power from diesel fuel is high; the supply of fresh water could also be an advantage. Essentially land-based, open-cycle or closed-cycle coupled with a desalinated process would be appropriate.

7.5.3. Advantages and Disadvantages

7.5.3.1. Advantages

OTEC is a clean, environmentally friendly renewable energy source that doesn't require burning fossil fuels, emitting a lot of greenhouse gases, or causing harmful air pollution. It sounds incredibly alluring. OTEC could contribute to lessening global collateral damage from an oil-dependent economy by assisting in reducing our reliance on fuels like petroleum. This could include averting conflicts over oil and reducing water pollution from tanker spills. Additionally, it might offer tropical island states without their own energy resources a very helpful source of power, essentially making them self-sufficient. Open-cycle OTEC can be helpful in producing clean, usable water from ocean water, as we've already discussed.

Moreover, OTEC can be used to create fuels like hydrogen. The electricity it produces can power an electrolysis plant, which would separate saltwater into hydrogen and oxygen. The hydrogen and oxygen would then be piped or bottled ashore and used to power devices like fuel cells in electric cars. Aquaculture (the controlled growth of fish and other marine food, such as algae) and air conditioning are further uses for the waste cooling water from an OTEC plant.

7.5.3.2. Disadvantages

The primary issue with OTEC is its relatively low efficiency. Any feasible heat engine must function at a temperature lower than 100% according to the laws of physics (the Carnot cycle); most operate much below this threshold. OTEC plants, which employ a relatively small temperature differential between their hot and cold fluids, have some of the lowest efficiency of all, often only a few percent. Because of this, OTEC plants must work extremely hard and pump enormous volumes of water to generate even small amounts of electricity, which poses two issues. First of all, it implies that a sizeable portion of the electricity produced—typically around a third—must be utilized for system operation, which includes pumping water in and out. Second, it suggests that OTEC plants require costly investments due to the relatively large scale of construction required. Significant environmental effects on shorelines, which are frequently the habitat of delicate, already endangered ecosystems like mangroves and coral reefs, could result from large-scale onshore OTEC plants (Chiles, 2009; Lewis, 2011) (Table 7.5).

Table 7.5. *Summary of the Main Oscillating Water Column (OWC) Devices in OTEC*

Project	Country	Position	Inst.	Pow-er	Status	Note
Kværner Brug's OWC	Toftes-fallen (Nor-way)	Fixed On-shore	1985	500 kW	Decom-missioned	Damaged by a severe storm in 1988. 29 MWh produced.

Project	Country	Position	Inst.	Power	Status	Note
Vizhinjam OWC	Trivandrum (Kerala, India)	Fixed Nearshore	1990	150 kW	Decommissioned in 2011	Production under the expectation, with high variability. Inactive for a long period.
Limpet	Islay (Scotland, UK)	Fixed Onshore	2000	500 kW	Decommissioned in 2012	High level of noise. The attenuation causes malfunctions.
Pico OWC	Pico Island (Azores, Portugal)	Fixed Onshore	1999	400 kW	The project concluded in 2018	Malfunction due to the Wells turbine and its supports.
Mutriku plant	Mutriku (Spain)	Fixed Onshore (harbor integrated)	2011	296 kW	Operating	Total electrical production 2.1 GWh (end of September 2020)
REWEC3	Civitavecchia (Italy)	Fixed Onshore (harbor integrated)	2016	2.5 MW	Operating	Annual production between 500 and 800 MWh/y
Yongsoo plant	Jeju Island (Republic of Korea)	Fixed Offshore (1 km to the coastline)	2016	500 kW	Operating	Adoption of two impulse turbines, connected to different generators
Wave Swell	King Island (Tasmania, Australia)	Fixed Nearshore (100 m to the coastline)	2019	200 kW	Operating	The device will test for a year

Project	**Country**	**Position**	**Inst.**	**Pow-er**	**Status**	**Note**
Backward Bent-Duct Buoy	Japan	Floating	1960–70	Con-cept		
Sloped IPS Buoy	Edin-burgh (UK)	Floating	1970–80	Con-cept		
Spar Buoy	Portugal	Floating	2000–15	Con-cept		
Mighty Whale	Gokasho Bay (Japan)	Floating Offshore (1.5 km to the coast-line)	1998	120 kW	Decom-missioned in 2002	

Source: Curto D, Franzitta V, Guercio A. Sea Wave Energy. A Review of the Current Technologies and Perspectives. Energies. 2021; 14(20):6604. https://doi.org/10.3390/en14206604.

7.6. IMPACT ON CARBON FOOTPRINT

Because ocean energy can produce electricity with low greenhouse gas emissions, it has a significant effect on lowering carbon emissions. Ocean energy technologies, as opposed to fossil fuels, use carbon-intensive fuels to capture natural marine processes like tides, waves, and ocean currents. Because of this essential feature, ocean energy can generate electricity while emitting almost no direct carbon dioxide during operation.

The lifecycle benefits of ocean energy are the main factor contributing to its reduced carbon footprint. During the production, installation, and maintenance stages, ocean energy systems—such as wave energy converters, tidal turbines, and ocean thermal energy systems—emit very little carbon dioxide. When compared to emissions from the burning of fossil fuels to generate electricity, these emissions are negligible. Ocean energy technologies, for instance, generally have lower lifecycle emissions than coal-fired power plants, which release between 800 and 1000 grams of CO_2-equivalent per kilowatt-hour (g CO_2e/kWh). However, depending on technology and implementation, lifecycle emissions from ocean energy systems can be as low as 10 to 30 g CO_2e/kWh.

Additionally, ocean energy offers a consistent and reliable power source that can replace generation based on fossil fuels, thereby lowering total carbon emissions. It is feasible to significantly reduce greenhouse gas emissions and support efforts to mitigate climate change by incorporating ocean energy into the energy mix.

Despite the low operational carbon footprint, optimizing the benefits requires careful consideration of the lifecycle emissions and potential environmental impacts. Ocean energy has the potential to be a major player in a low-carbon energy future, assisting initiatives to reduce global carbon footprints and promote sustainable energy practices, thanks to continuous technological advancements and increased efficiency.

7.6.1. Potential Contributions to Emissions Reduction

Ocean energy holds significant potential for reducing greenhouse gas emissions and contributing to global climate goals. By harnessing the power of ocean tides, waves, currents, and thermal gradients, ocean energy can offer a low-carbon alternative to fossil fuel-based electricity generation.

1. **Consistent and Predictable Power:** Ocean energy sources, such as tidal and wave power, provide a stable and reliable supply of energy. Unlike wind and solar power, which are intermittent, tidal movements and ocean currents are highly predictable. This reliability allows ocean energy to consistently displace fossil fuel generation, leading to significant reductions in carbon emissions. For example, tidal power projects can contribute to consistent baseload power, reducing the need for coal or natural gas backup.
2. **Low Operational Emissions:** Ocean energy systems generally have very low operational carbon footprints. During operation, these technologies do not emit greenhouse gases, unlike fossil fuel power plants. The primary emissions associated with ocean energy come from the manufacturing, installation, and maintenance of the equipment. However, these lifecycle emissions are considerably lower compared to those from fossil fuel-based power sources.
3. **High Energy Density:** Ocean energy technologies, such as ocean current and wave energy converters, can generate substantial amounts of energy in localized areas, which can offset significant quantities of fossil fuel use. For example, large-scale wave and tidal projects can produce high-density energy, contributing substantially to regional and national energy supplies.

4. **Complementary to Other Renewables:** Ocean energy can complement other renewable energy sources by providing a steady energy output that balances the variability of wind and solar power. Integrating ocean energy with existing renewable energy systems can enhance overall grid stability and further reduce reliance on fossil fuels.
5. **Climate Change Mitigation:** By reducing dependence on carbon-intensive energy sources, ocean energy supports broader climate change mitigation strategies. As technology advances and costs decrease, the widespread adoption of ocean energy can play a crucial role in achieving long-term emissions reduction targets.

Overall, ocean energy's ability to provide reliable, low-carbon power makes it a valuable contributor to reducing global greenhouse gas emissions and advancing toward a sustainable energy future.

7.6.2. Economic and Environmental Considerations: Cost, Feasibility, and Ecological Impacts

When evaluating ocean energy, it's crucial to consider the economic and environmental factors that influence its deployment and impact.

These include cost, feasibility, and ecological impacts:

1. **Cost:** When compared to other renewable energy sources, the initial investment for ocean energy technologies is comparatively higher. This is because these technologies operate in a complex and frequently harsh marine environment, which drives up installation and maintenance costs. Advanced materials and engineering are needed for technologies like tidal turbines, wave energy converters, and ocean thermal energy systems because they must withstand strong marine forces and corrosive saltwater. But costs should go down as technology develops and economies of scale become apparent. Long-term advantages of ocean energy include low operating costs and no fuel expenses, which can eventually make it a financially viable choice.
2. **Feasibility:** The availability of resources, the state of technology, and the infrastructure needs all affect how feasible ocean energy projects are. Because ocean energy resources are site-specific, their effective application is frequently restricted to regions with substantial thermal gradients, strong tidal currents, or high wave energy. The distribution and scalability of ocean energy projects may be impacted by this geographical restriction. Another aspect is

technological maturity; although some ocean energy technologies are still in the early phases of development or commercial application, continued research and pilot projects are progressively proving their viability and potential for broader application.

3. **Ecological Impacts**: Projects involving ocean energy may have diverse ecological effects, which call for careful management through environmental assessments. These effects include physical interactions with marine organisms, changes in sediment transport patterns, and potential disruptions to marine habitats and wildlife. The local fishing industry and ecosystems may be impacted by tidal and wave energy installations. Comprehensive environmental impact assessments, efficient monitoring programs, and design changes are necessary to lessen these effects. Efficient energy production coupled with minimal ecological disturbance is essential to weighing the advantages and disadvantages of ocean energy. Ocean energy has a lot to offer in terms of cutting carbon emissions and producing dependable renewable energy, but successful implementation and sustainable development depend on resolving financial obstacles, determining viability, and controlling ecological effects (Chiles, 2009; Lewis, 2011).

REFERENCES

1. Chiles, J. (2009). The other renewable energy. *Invention and Technology, 23*(4), 24–35.
2. Eldred, M., Landherr, A., & Chen, I. C. (2010). Comparison of aluminum alloys and manufacturing processes based on corrosion performance for use in OTEC heat exchangers. In *Offshore Technology Conference 2010 (OTC 2010)*. Curran Associates, Inc.
3. Lewis, A., et al. (2011). *IPCC: Special Report on Renewable Energy Sources and Climate Change Mitigation.*
4. Rocheleau, G. J., & Grandelli, P. (2011). Physical and biological modeling of a 100-megawatt Ocean Thermal Energy Conversion discharge plume.
5. Thomas, D. (1999). A brief history of OTEC research at NELHA. *NELHA*.
6. Vega, L. A. (2010). *Economics of Ocean Thermal Energy Conversion* (PDF). National Marine Renewable Energy Center at the University of Hawaii.

Chapter 8

Integration and Storage of Renewable Energy

Contents

INTRODUCTION

Optimizing the efficiency and dependability of contemporary power systems depends heavily on the integration and storage of renewable energy. Effective integration and storage solutions are required as the share of renewable energy sources, such as wind and solar, increases due to their intermittent nature, which is dependent on the weather and time of day. Real-time supply and demand balancing is made easier by advanced grid management technologies like demand response systems and smart grids. To improve grid stability and handle variable renewable inputs, smart grids monitor and control energy flows using digital communication technologies.

Technologies for energy storage are essential for mitigating the unpredictability of renewable energy sources. Lithium-ion and developing solid-state batteries, among others, provide scalable means of storing excess energy produced during periods of peak production. A constant supply of power can then be ensured by deploying these stored energies during times of low production or high demand. Another well-known technique offering a sizable storage capacity is pumped hydro storage, which entails pumping water to a higher elevation during times of excess energy and releasing it to generate power when needed.

Enhancing grid infrastructure to handle a variety of energy sources is another aspect of integrating renewable energy, in addition to storage. This entails creating microgrids that can function separately or in tandem with the main grid, as well as modernizing transmission lines and regional connectivity. Green hydrogen, which is created from surplus renewable energy, is showing promise as a long-term energy storage option. It allows energy to be chemically stored and used in a variety of applications, such as industry and transportation. All things considered, optimizing the advantages of clean energy, maintaining grid reliability, and accelerating the shift to a sustainable energy future depend on the efficient integration and storage of renewable energy. These solutions help to create a more flexible and resilient energy system by addressing the issues of intermittency and optimizing resource use (Akhil et al., 2013; Bhatnagar et al., 2013).

8.1. OVERVIEW OF ENERGY STORAGE

OE's Energy Storage Program conducts research and development on a wide range of storage technologies because energy storage technology can be applied to many different areas with varying power and energy requirements. Batteries (both conventional and advanced), electrochemical capacitors, flywheels, power electronics, control systems, and software tools for storage optimization and sizing are just a few examples of this wide range of technologies. Numerous of the Energy Storage Program's projects involve significant cost sharing, and it collaborates closely with industry partners. To design, procure, install, and commission significant innovative storage installations that can reach several megawatts in size, the Program also works with utilities and state energy organizations like the Massachusetts Clean Energy Center (MASS CEC), California Energy Commission, Oregon DOE, Vermont, Hawaii, Washington, and New York State Energy Research and Development Authority (NYSERDA), to mention a few. Additionally, it facilitates technical assessments of operating systems and ES system components as well as analytical research on the technical and financial performance of storage technologies.

Enhanced energy storage can provide multiple benefits to both the power industry and its customers. Among these benefits are:

1. Improved power quality and the reliable delivery of electricity to customers.
2. Improved stability and reliability of transmission and distribution systems.
3. Increased use of existing equipment, thereby deferring or eliminating costly upgrades.
4. Improved availability and increased market value of distributed generation sources.
5. Improved value of renewable energy generation.
6. Cost reductions through capacity and transmission payment deferral.

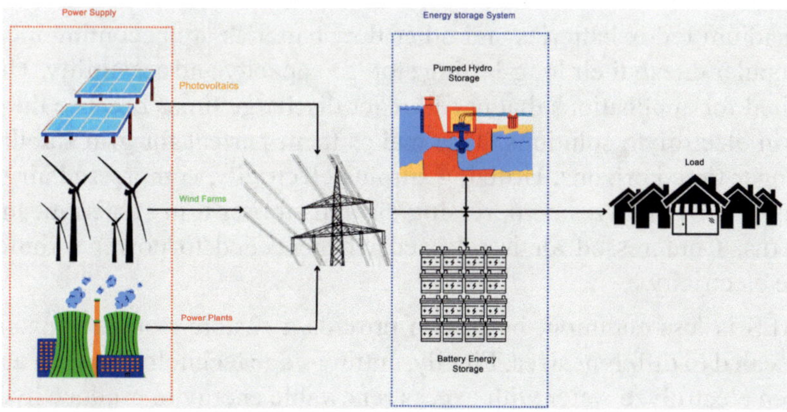

Figure 8.1. *An overview of energy storage systems.*

Source: Sayed ET, Olabi AG, Alami AH, Radwan A, Mdallal A, Rezk A, Abdelkareem MA. Renewable Energy and Energy Storage Systems. Energies. 2023; 16(3):1415. https://doi.org/10.3390/en16031415.

The energy storage program also seeks to improve energy storage density by conducting research into advanced electrolytes for flow batteries, development of low temperature Na batteries, along with and nano-structured electrodes with improved electrochemical properties. In Power Electronics, research into new high-voltage, high power, high frequency, wide-band-gap materials such as silicon-carbide and gallium-nitride is underway (Figure 8.1).

8.1.1. Energy Storage Technologies

For the best possible utilization of renewable energy sources—which are frequently erratic and intermittent—energy storage technologies are essential. Lithium-ion batteries are one of the most popular storage technologies because of their high energy density, extended cycle life, and quick response times. To balance supply and demand, these batteries are being utilized more frequently in both residential and commercial applications to store extra solar and wind energy for later use. Emerging technology known as solid-state batteries holds the potential to increase energy storage's viability and efficiency by offering even higher energy densities and safety enhancements. An additional well-established method for storing energy on a bigger scale is pumped hydro storage.

It functions by pumping water from a lower reservoir to a higher elevation using extra electricity. The water that has been stored is released to flow back down through turbines, producing electricity, during times of high demand or low renewable output. This approach offers a large amount of storage capacity and reacts fast to variations in energy consumption.

Vanadium redox batteries and other flow batteries are becoming more and more popular due to their long-lasting storage capacity and scalability. They are best suited for applications that need longer discharge times because they store energy in electrolyte solutions. This makes them perfect for grid stabilization over longer time horizons. Utilizing surplus electricity, compressed air energy storage (CAES) involves compressing air and storing it in subterranean tanks or caverns. Compressed air is released when needed to power turbines and produce electricity.

CAES is less common, but it can provide a sizable storage capacity and can be scaled to different sizes. Finally, cutting-edge technologies such as green hydrogen electrolyze water with excess renewable energy to produce hydrogen that can be stored, burned as fuel, or transformed back into electricity. This method supports industries and transportation in addition to the electrical sector by offering long-term storage options. All things considered, these energy storage technologies play a critical role in augmenting the dependability and adaptability of renewable energy systems, facilitating a more seamless shift towards a sustainable energy profile (Akhil et al., 2013; Bhatnagar et al., 2013).

8.1.1.1. A Wide Array of Storage Technologies Have Been Developed so that the Grid Can Meet Everyday Energy Needs

Since the discovery of electricity, we have sought effective methods to store that energy for use on demand. Over the last century, the energy storage industry has continued to evolve, adapt, and innovate in response to changing energy requirements and advances in technology.

Energy storage systems provide a wide array of technological approaches to managing our power supply to create a more resilient energy infrastructure and bring cost savings to utilities and consumers. To help understand the diverse approaches currently being deployed around the world, we have divided them into five main categories:

1. **Batteries** – a range of electrochemical storage solutions, including advanced chemistry batteries, flow batteries, and capacitors.
2. **Thermal** – capturing heat and cold to create energy on demand or offset energy needs.
3. **Mechanical Storage** – other innovative technologies to harness kinetic or gravitational energy to store electricity.
4. **Hydrogen** – excess electricity generation can be converted into hydrogen via electrolysis and stored.
5. **Pumped Hydropower** – creating large-scale reservoirs of energy with water.

8.1.2. Benefits of Energy Storage

The investment in energy storage evolves with our grid, creating long-term benefit and reliability for years to come. Energy storage is a critical hub for the entire grid, augmenting resources from wind, solar, and hydro, to nuclear and fossil fuels, to demand side resources and system efficiency assets. It can act as a generation, transmission or distribution asset – sometimes in a single asset.

Ultimately, storage is an enabling technology. It can save consumers money, improve reliability and resilience, integrate generation sources, and help reduce environmental impacts (Table 8.1).

Table 8.1. *Renewable Energy Installed Capacities and Electricity Production in the World*

Continent	Solar Energy		Wind Energy		Bioenergy		Hydropower	
	Capacity (MW)	Energy (GWh)	Capacity (MW)	Energy (GWh)	Capacity (MW)	Energy (GWh)	Capacity (MW)	Energy (GWh)
Asia	485,948	447,985	385,393	555,824	56,969	211,827	594,267	1,927,807
Africa	11,393	17,037	7334	10,557	1785	3203	37,677	141,437
Europe	187,360	167,605	220,760	488,412	41,712	206,760	224,393	608,207
North America	105,881	137,703	154,733	397,157	16,956	72,683	198,026	722,078
South America	19,649	22,126	29,754	79,601	18,484	71,882	178,033	672,569
Oceania	23,242	21,548	9827	22,760	1104	4285	14,498	41,314
Total in the World	854,795	843,855	823,484	1,588,586	143,195	583,775	1,360,502	4,476,230

Source: Sayed ET, Olabi AG, Alami AH, Radwan A, Mdallal A, Rezk A, Abdelkareem MA. Renewable Energy and Energy Storage Systems. Energies. 2023; 16(3):1415. https://doi.org/10.3390/en16031415.

8.1.2.1. Save Money

Energy storage can save operational costs in powering the grid, as well as save money for electricity consumers who install energy storage in their homes and businesses. Energy storage can reduce the cost to provide frequency regulation and spinning reserve services, as well as offset the costs to consumers by storing low-cost energy and using it later, during peak periods at higher electricity rates.

By using energy storage during brief outages, businesses can avoid costly disruptions and continue normal operations. Residents can save themselves from lost food and medicines, and the inconvenience of not having electricity. And there is an option for both businesses and residential consumers to participate in demand response programs when available.

8.1.2.2. Improve Reliability and Resilience

Energy storage can provide backup power during disruptions. The same concept that applies to backup power for an individual device (e.g., a smoke alarm that plugs into a home but also has battery backup), can be scaled up to an entire building or even the grid at large.

Storage provides flexibility for the grid, to ensure uninterrupted power to consumers, whenever, and wherever they need it. This flexibility is critical to both reliability and resilience. As the cost of outages continues to rise, the value of enhanced reliability and improvements in resilience also increases.

8.1.2.3. Integrate Diverse Resources

By storing excess energy when the sun is shining and the wind is blowing and delivering it when the opposite is true, energy storage can even out the delivery of variable or intermittent resources like solar and wind.

However, storage can also help with the effective delivery of power for baseload, inflexible resources. Energy storage can precisely match load by injecting or extracting electricity as needed, wherever, and whenever it's needed, when demand fluctuates quickly and flexibility is needed.

One technological enabler is energy storage. Energy storage is possible even in the absence of the sun or wind. Energy storage may be available when demand changes and baseload resources are unable to respond promptly.

8.1.2.4. Reduce Environmental Impacts

To put it simply, energy storage makes it possible to store electricity for use at a later time, location, and need. As a result, the electric grid gains capabilities and efficiencies, including the capacity to lower greenhouse gas (GHG) emissions.

Energy storage can assist in integrating more solar, wind, and distributed energy resources by giving the grid more flexibility. It can also reduce the need to construct new peak power plants that emit pollution and increase the efficiency of the grid by raising the capacity factor of currently available resources.

Energy storage makes it easier and more reliable for our energy supply mix to change as it becomes more sustainable by incorporating low- and no-carbon resources (Akhil et al., 2013; Bhatnagar et al., 2013).

8.1.3. Grid Integration

The grid integration of renewable energy is a complicated but necessary procedure for updating power systems and guaranteeing a consistent, sustainable energy supply. This is another crucial element. Due to their inherent intermittency brought on by variations in the weather and time of day, renewable energy sources like wind and solar are becoming more and more common. This poses challenges for maintaining grid stability and reliability. So, to fully utilize these clean energy sources, effective grid integration strategies are essential.

To effectively integrate grids, advanced grid management technologies must be developed. Smart grids improve the capacity to track and manage energy flows in real-time by including sensors and digital communication tools. By enabling dynamic responses to variations in energy supply and demand, these technologies enhance the resilience and efficiency of the grid. Improved forecasting of renewable energy generation and demand patterns is made possible by smart grids, which aids operators in managing variability and lowers the chance of brownouts or blackouts command response, which modifies power consumption habits in response to grid circumstances. Demand response programs aid in the optimization of the use of renewable resources, the reduction of peak loads, and the balancing of supply and demand by providing incentives to consumers to shift their energy consumption to times of high renewable generation or low demand. This adaptability is essential for successfully integrating variable renewable energy sources.

Technologies for energy storage are essential for mitigating the intermittent nature of renewable energy sources. Lithium-ion and newer solid-state batteries, among others, store excess energy produced during periods of high production and release it during periods of low generation. Pumped hydro storage offers significant capacity and quick response by using extra energy to pump water to higher altitudes for later use. For extended periods of time, additional storage alternatives like compressed air energy storage and flow batteries also help to balance the grid. Infrastructure on the grid needs to be expanded and upgraded to accommodate the growing percentage of renewable energy. Rebuilding regional connections, creating decentralized microgrids, and updating transmission lines can all help the grid handle a wider range of geographically distributed energy sources. By enabling the sharing of renewable energy over larger regions, interconnections lessen the effects of regional variations and improve grid stability overall. Lastly, careful planning and coordination amongst multiple stakeholders, such as utilities, regulators, and technology providers, are necessary for the integration of renewable energy.

To successfully transition to a renewable-powered grid, policies, and regulations that support grid modernization, encourage investment in clean energy technologies, and facilitate collaboration are essential. To summarize, the integration of renewable energy into the grid requires a blend of cutting-edge technologies, upgraded infrastructure, and strategic planning.

Through the resolution of intermittency issues and the enhancement of grid management, these endeavors bolster the dependable and effective utilization of renewable energy sources, propelling the shift towards a more robust and sustainable power infrastructure (Akhil et al., 2013; Bhatnagar et al., 2013).

8.1.4. Challenges and Solutions for Integrating Renewable Energy into Existing Grids

Integrating renewable energy into existing grids presents several challenges, but there are effective solutions to address these issues and enhance grid reliability and efficiency.

Here's a detailed look at the primary challenges and their corresponding solutions:

8.1.4.1. Challenges

1. Intermittency and Variability
    - **Description:** Renewable energy sources like wind and solar are inherently variable, with their output fluctuating based on weather conditions and time of day. This intermittency can lead to mismatches between energy supply and demand.
    - **Solution: Energy Storage Systems** such as batteries, pumped hydro, and flow batteries can store excess energy during periods of high generation and release it when generation is low. Demand Response Programs can also help by shifting energy use to match periods of high renewable output, thereby balancing supply and demand.
2. Grid Stability and Reliability
    - **Description:** The integration of variable renewable energy

sources can affect grid stability, causing frequency and voltage fluctuations. This variability requires careful management to avoid grid imbalances or outages.

- **Solution: Advanced Grid Management Technologies** including smart grids and real-time monitoring systems can improve grid stability by providing real-time data and enabling dynamic responses to fluctuations. Flexible Resources such as fast-responding gas turbines or virtual power plants can also help stabilize the grid.

3. Infrastructure Limitations
   - **Description:** Many existing power grids were designed for centralized, fossil-fuel-based generation and may not be equipped to handle the decentralized and variable nature of renewables. This can lead to bottlenecks and inefficiencies.
   - **Solution: Grid Modernization** involves upgrading transmission lines, expanding interconnections, and developing decentralized microgrids to better accommodate renewable energy. Smart Grid Technologies can enhance the management of distributed energy resources and improve overall grid performance.
4. Regulatory and Market Challenges
   - **Description:** Current energy markets and regulatory frameworks may not fully support or incentivize the integration of renewable energy. This can create barriers to investment and deployment of renewable technologies.
   - **Solution: Policy and Regulatory Reforms** are needed to support renewable energy integration. This includes updating market structures to value the benefits of renewables, implementing renewable portfolio standards, and providing incentives for clean energy investments. Cross-Sector Collaboration between governments, utilities, and industry stakeholders can facilitate these changes.
5. Economic and Financial Barriers
   - **Description:** High upfront costs for renewable energy projects, grid upgrades, and storage systems can be a significant barrier, especially in regions with limited financial resources.
   - **Solution:** Innovative Financing Models such as green bonds, public-private partnerships, and community-based funding can help lower financial barriers. Additionally, Economies

of Scale achieved through larger projects and technological advancements can reduce costs over time.

Addressing these challenges requires a multifaceted approach involving technological advancements, infrastructure improvements, regulatory changes, and financial innovation. By leveraging energy storage, enhancing grid management, modernizing infrastructure, and reforming policies, the integration of renewable energy into existing grids can be achieved more effectively, leading to a more resilient, reliable, and sustainable energy system.

8.2. ROLE OF ENERGY STORAGE

The role that electricity and heat storage systems play in the steady decline of fossil fuels and the quick ascent of renewable energy sources. Climate change, clean energy initiatives, energy transition, and sustainable energy systems are all responsible for the decline in fossil fuel consumption and the rise in renewable resources. The rapid expansion of renewable energy sources presents new operational and design difficulties in the transition to the goal of 100% renewable energy. Systems for storing energy can facilitate the ride-through of energy from fossil fuels to renewable energy sources. Artificial photosynthesis and nuclear fusion represent the ultimate holy grails of sustainable clean energy. To control the food supply chain and potentially serve as a model for the energy transition from fossil fuels to renewables, plants absorb light and store it as chemical energy. Solar and wind energy sources' periodic nature can be explained by devices that store heat and electricity. In regions with below-freezing temperatures, solar thermal systems are another practical option for heating water and space (Akhil et al., 2013; Bhatnagar et al., 2013).

8.2.1. Trends in Energy Storage

8.2.1.1. Where Is the Energy Storage Industry Today?

Global power systems have concentrated on producing, transmitting, and distributing electricity as a real-time commodity for almost a century. Physics demands that, even in the face of load variability on timescales ranging from multiyear trends to sub-second disturbances, electricity generation always be in real-time balance with the load. Grid operators are employing new techniques to keep this balance in light of the growing importance of variable generation from wind and solar, the retirement of fossil fuel-based generation, and shifting consumer demand profiles.

When used properly, energy storage provides system operators with a quick-response and adaptable resource to efficiently handle load and generation variability. The cost of battery energy storage has recently dropped dramatically, following the learning curves observed in wind and solar power generation over the previous 10 years. Many believe that battery energy storage will play an increasingly important—if not crucial—role in the future grid due to their decreasing costs. 78% of utility professionals expect utility-scale storage to increase moderately or significantly on their systems, according to Utility Dive's State of the Electric Utility 2017 survey. In contrast to other utility-scale energy technologies, battery energy storage is still a relatively new and expensive technology. Even though 2016 turned out to be a record year for the deployment of battery energy storage, there were only 221 MW of installations nationwide. Even though energy storage has a bright future, it is still a support system for the electrical grid. In the United States, where there is a peak demand of almost 840 GW, energy storage only makes up 24 GW of capacity, even when big pumped hydro plants are taken into account. Less than 1 GW of battery storage and 0.8 GW of thermal storage are available at the moment. Other electrical grids across the world can be described similarly. Pumped hydro plants account for more than 95% of the world's current energy storage capacity. These plants have historically been constructed as part of the nuclear power expansion in the 1970s and early 1980s, with very little new construction taking place in the last 20 years (Table 8.2).

The technology of battery energy storage is still developing. The state of energy storage technology today is limited despite recent rapid advancements and development. These include high costs in terms of both kW and kWh, lower-than-desired rates of charge and discharge, life-cycle limitations, and worries regarding the safety of battery energy storage systems. Furthermore, there may not be a "one size fits all" solution due to the vast range of applications for energy storage, which call for significantly different levels of power, energy, cost, cycle life, size, and flexibility. Rather, the industry will require a range of energy storage technologies, from long-duration storage required for energy applications to high-power, short-duration batteries needed for balancing applications.

Table 8.2. *Renewable Energy Installed Capacities and Electricity Production in the Most Contributed Countries*

Conti-nent	**Solar Energy**		**Wind Energy**		**Bioenergy**		**Hydropower**	
	Capacity (MW)	**Energy (GWh)**	**Capacity (MW)**	**Energy (GWh)**	**Capacity (MW)**	**Energy (GWh)**	**Ca-pacity (MW)**	**Energy (GWh)**
United States	95,209	119,329	132,738	341,818	13,574	60,269	101,894	308,213
China	306,973	261,659	328,973	467,037	29,753	98,978	390,920	1,355,210
India	49,684	54,666	40,067	63,522	10,592	21,987	51,565	164,678
Cana-da	3630	4846	14,304	35,638	2416	10,094	82,740	386,617
Ger-many	58,728	48,641	63,865	132,102	10,439	50,858	10,739	24,876
Japan	74,191	79,087	4467	8970	4592	27,995	50,019	87,548

Source: Sayed ET, Olabi AG, Alami AH, Radwan A, Mdallal A, Rezk A, Abdelkareem MA. Renewable Energy and Energy Storage Systems. Energies. 2023; 16(3):1415. https://doi.org/10.3390/en16031415.

8.2.2. Rise of Lithium Ion and the Competitive Landscape of Storage Technologies

The majority of battery energy storage applications currently use lithium-ion (Li-ion) batteries because the technology has come a long way in the last 20 years, both in terms of cost and functionality. Li-ion chemistry was responsible for at least 97% of grid-scale battery energy storage deployment in 2016, according to GTM Research. Although Li-ion battery storage systems have experienced the fastest rate of cost reduction and deployment expansion, other cell chemistries' battery systems are also making progress. The scaling up of new battery technologies, such as flow batteries, alkaline rechargeable batteries, and lower temperature sodium batteries (NaS and $NaNiCl_2$), is still progressing significantly. While NaS battery technology has been proven through decades of production and application, newer technologies like flow batteries and alkaline rechargeable storage systems need to show that they can satisfy the reliability and cost requirements of electric utilities. Furthermore, unlike Li-ion batteries, more recent storage technologies lack access to other sizable markets, like electric cars, which could have covered the upfront costs and learning cycles required to scale up to high-volume production. The use of battery-based energy storage has gained traction in the market during the past few years. In actuality, 2016 marked the first year in which the yearly deployment of battery-based storage exceeded one gigawatt-hour (Table 8.3).

Table 8.3. *Energy Storage Systems Comparison*

Technology	Power Rating (MW)	Cycle Efficiency (%)	Lifetime (Years)	LCOE ($/kWh)
Compressed Air	110–1000	42–54	20–40	2–120
Pumped Hydro Storage	30–5000	70–87	40–60	5–100
Thermal Energy Storage	0.1–300	30–60	20–30	3–60
Lead Acid Batteries	0–40	63–90	5–15	50–400
Li-ion Batteries	0–100	75–97	14–16	600–3800
Flywheels	0.25–20	90–95	15–20	1000–14,000
Supercapacitors	0–0.3	84–97	10–30	300–2000
Fuel Cells	<58.5	20–66	~20	2–15

Technology	Power Rating (MW)	Cycle Efficiency (%)	Lifetime (Years)	LCOE ($/kWh)
Super magnetic Conducting Energy Storage	0.1–10	95–98	20–30	500–72,000

Source: Sayed ET, Olabi AG, Alami AH, Radwan A, Mdallal A, Rezk A, Abdelkareem MA. Renewable Energy and Energy Storage Systems. Energies. 2023; 16(3):1415. https://doi.org/10.3390/en16031415.

In the vast majority of application markets, electrochemical energy storage (through battery storage systems) offers the most flexibility in terms of energy and power capacity (MW and MWh ratings); this includes large projects to support TandD infrastructure deferrals, 50–100 MWhpeaker replacements, and kWh class behind-the-meter applications. The majority of market studies indicate that deployments across all market segments will increase at a rapid pace, indicating the rapid expansion of the U.S. and global markets for energy storage technologies. The global grid storage market is predicted to grow by more than 10 times by 2025, with almost 80% of that growth occurring outside of the United States, according to Bloomberg New Energy Finance.

This adoption can be accelerated by more cost reductions and better market mechanisms that fairly monetize the advantages of energy storage. The marked decline in Li-ion battery prices is directly responsible for the acceleration of the development of Li-ion battery-based grid storage systems. Cell level costs have decreased to less than $200/kWh thanks to a large increase in Li-ion manufacturing production capacity; however, more cost reductions may be more difficult to attain than those that have already been made. The battery system's capital cost is merely one component of the total cost. In addition to the storage device, other expenses include those related to integration, power conditioning, battery management systems, and soft costs related to system installation. Because of this, even though the cost of battery packs has significantly decreased, similar cost reductions will be necessary for the other parts and balance of plants that make up an energy storage system. It should be mentioned that some of the related expenses go toward more established technologies like transformers and inverters, as well as related infrastructure like site development. Depending on the power-to-energy ratio required for the application, the DC subsystem (which primarily consists of battery cells) accounted for between 35 and 70% of the total system cost at the end of 2016. The wide variation stems from the kind of arrangement, where low-power, high-energy systems (used for firm capacity and energy shifting) are utilized more frequently than high-power, low-energy systems (used for frequency regulation).

The majority of cost savings to date have resulted from advancements in battery cell technology. Innovations in power electronics and power conversion systems, sensors, software, and control systems will need to be incorporated to further reduce costs. The power component of battery energy storage systems still needs to scale, even though the energy components (Li-ion battery cells) are starting to become "commoditized," partly as a result of improvements in the production of electric vehicles. Custom or semi-custom balance of plant is present in the majority of battery installations. The balance of plant costs for battery energy storage is higher than those for solar technology, even though the underlying technologies—such as inverters—may be more advanced. This is because full integration and controls are still in their infancy. Additionally, there are more opportunities to cut soft costs through system-level standardization, enhanced plant layouts, and expedited construction, permitting, and startup costs. The safety, dependability, and operational efficiency of energy storage systems will continue to improve with increased deployment and operating experience.

This will result in more consistent maintenance plans and enhanced operational efficiency, which will encourage utilities and grid operators to use energy storage more widely. Storage has the potential to play a bigger part in the grid infrastructure. However, storage cannot be implemented at the necessary scale due to the current cost structure. The majority of large-scale deployments have been restricted to niche markets where storage ancillary services have been sufficiently profitable or where government regulations require a minimum level of storage deployment. Storage, for instance, is being used in regulatory services where there are market mechanisms in place to compensate for the quicker response that storage assets can offer.

In the years since the PJM Regulation Market opened to storage competition in 2011, more than 300 MW of new energy storage systems have been installed. As a result, the price of regulation services was cut in half. Energy storage can perform faster or better than conventional resources for a variety of reasons, but there are no established market mechanisms for these uses. In contrast to frequency regulation, bulk applications for other services—like generation optimization, resiliency, or infrastructure deferrals—are currently not financially feasible due to the relatively high cost of existing technology. However, these services will require new technologies and operating procedures as wind and solar integration rises and conventional thermal generators are retired. This opens up new possibilities for energy storage, but before many of these services can be made profitable, markets, and utility planning need to adjust. There are currently niche applications with attractive economics, and as capital costs continue to drop and the penetration of renewable energy sources rises, these applications should become more widespread (Akhil et al., 2013; Bhatnagar et al., 2013).

8.2.3. The Multiple Value Streams of Energy Storage

8.2.3.1. The Costs Are Coming Down, but Where Are the Markets?

Batteries and other emerging technologies will play a variety of roles and enter a variety of markets as long as their cost and performance remain low. Deployment to short-duration devices (less than one hour of stored energy) was largely restricted by the historically high prices of battery energy capacity, with a focus on the frequency regulation market, which usually has the highest clearing price of the ancillary service markets in the US. About 2.5 GW is the total market for frequency regulation in US RTOs and ISOs, which supply roughly two-thirds of the country's demand. Storage developers are looking beyond this small market because storage will face increasing competition in this market from demand response, curtailed wind and solar, and other flexible resources.

Short-duration storage can also be used for fast frequency response and spinning contingency reserves. Although most regions do not yet have a well-established market for fast frequency response, there are a number of active proposals and pilot projects being worked on in locations such as Hawaii, ERCOT, CAISO, and Ireland. Even though the market for short-duration storage is smaller than that of the power system overall, these early deployments will contribute to lowering the cost of storage's power and energy components. Additionally, they will give utilities and system operators invaluable experience with storage technologies, enabling them to play an even bigger part in the generation portfolio as a whole.

Beyond uses for temporary storage, storage has a far wider potential to serve as a capacity resource. Older peaking capacity will need to be replaced even in the absence of load growth; in the US alone, there are more than 100 GW of peaking capacity that is over 40 years old. Some of this retiring peaking capacity could be replaced by batteries that have a capacity of several hours. The need for a full replacement of the current peaking capacity will be lessened, nevertheless, due to factors such as demand response, energy efficiency, and the trend of most industrialized countries using less electricity.

On the basis of capital costs alone, storage will find it challenging to compete with conventional peaking resources like combustion turbines. At this point, it's critical to recognize and capitalize on the value of "stacked services." When capacity value is combined with the flexible dispatch and siting benefits of energy storage, even in the absence of potential opportunities for residual value in ancillary service markets, energy storage can have a higher initial cost than a new combustion turbine (CT) but a lower lifetime cost. Energy storage can be used to arbitrage price differences, prevent thermal plant starts, and provide long-duration operating reserves. One example of this is the flexi- ramp produce, which is currently being introduced or proposed in several US markets. An additional advantage is the possible flexibility in storage location. By charging during times of low losses, siting storage closer to load can prevent losses in the transmission and distribution (TandD) network, thereby improving its system-wide benefit measurably. It can also postpone investments in transmission and distribution assets.

8.2.4. Increased Renewable Energy Will Change the Economics for Battery Storage

The use of renewable energy could hasten the deployment of storage and open up new business opportunities. Incorporating renewable energy sources involves a complicated and sometimes contentious role for storage. Studies on the integration of renewable energy have not yet revealed an innate "need" for storage to use renewable energy at penetration rates of up to roughly 35% annually. According to other research, there are probably more affordable ways to deal with high ramp rates and the inevitable reduction in renewable energy that happens when supply is high and demand is low (at least given current storage costs). Significant amounts of solar and wind power have been combined in many areas without the need for battery energy storage. For instance, ERCOT achieved 50% instantaneous penetration in 2016 and generated more than 15% of its annual load from renewable sources. SPP was higher than 17% yearly and 52% instantaneous. Ireland has seen increases of more than 20% yearly and 60% instantaneous. As a result, the sector has demonstrated that battery energy storage is not necessary to achieve the levels of renewable integration that exist today. But as solar and wind energy become more and more prevalent, there will probably come a time when adding battery energy storage will be the most economical way to integrate renewables even further. This turning point is very system-specific and is determined by the price of fuel, the need for ancillary services, the region's generation mix, storage costs, and fuel prices. It is possible to draw the conclusion that there is room for significant synergies between storage and renewable energy, which will make both technologies more cost-competitive, by looking at how renewable energy affects the grid.

Some see energy storage as a natural counterbalance to the inherent unpredictability of the resource. Despite the fact that energy storage can discharge during cloudy or calm days and charge during windy or sunny days, industry experience has shown that most power grids do not currently require energy shifting of renewable resources from one time period to another. Currently, the majority of battery energy storage's value to renewable integration lies in its ability to control the power grid's short-term variability. As system load varies along with wind and solar generation, short-duration charging and discharging can offer load following and frequency regulation. In the event of severe disruptions or emergencies, grid-scale energy storage also offers a quick response for quickly injecting or withdrawing power to maintain grid stability (Akhil et al., 2013; Bhatnagar et al., 2013) (Table 8.3).

Table 8.4. *Largest Energy Storage Systems Projects Around the Globe*

Power Plant Name	Technology	Country	Year	Installed Capacity (MW)
Three Gorges Dam	Hydroelectric Power	China	2003	22,500
Itaipu Dam	Hydroelectric Power	Brazil and Paraguay	1984	14,000
Ouarzazate Solar Power Station	Thermal Storage (Molten Satl)	Morocco	2018	Total of 7325 MW_t NOOR I (1200) MW_t NOOR II (3500) MW_t NOOR III (2625) MW_t
Manatee Energy Storage Center	Batteries	United States	2021	409
Moss Landing Vistra Battery	Lithium-Ion Batteries	United States	2020	400

Source: Sayed ET, Olabi AG, Alami AH, Radwan A, Mdallal A, Rezk A, Abdelkareem MA. Renewable Energy and Energy Storage Systems. Energies. 2023; 16(3):1415. https://doi.org/10.3390/en16031415.

Increased use of renewable energy sources will eventually alter the storage market in three different ways. First, as solar PV penetration rises, the net demand profile will shift, making loads "peakier," and this will create more opportunities for storage with shorter durations—less than eight hours—to meet peak demand requirements.

Pumped storage makes up the majority of storage that is used both domestically and abroad. To ensure consistent peaking capacity, these power plants usually have storage for more than eight hours. Because PV produces shorter peaks, more energy storage capacity—6, 4, or even less hours—can be stored to achieve comparable reliability levels, making battery energy storage a more cost-effective peaking resource. Second, off-peak prices are typically suppressed by PV and wind, and arbitrageable low-cost baseload energy resources (coal and combined cycle generators with the highest efficiency) are more readily available. This happens even in the absence of appreciably reduced renewable energy.

Lastly, in an effort to cut costs, hybrid renewable and storage projects are using shared components. One way to share the use of the inverter, transformer, and system interconnection in a photovoltaic system is to deploy storage on the DC side of the interconnection. This is not to say that storage and particular renewable projects should be "coupled." While it might be practical to consider adding storage to a standalone renewable project to produce a "firm" renewable generator, such applications would be extremely inefficient because they would largely eliminate the value of value stacking—that is, the capacity of storage to charge from various sources to guarantee that stored energy is available whenever the grid needs it.

The deployment of storage would increase the economic deployment of renewable energy sources during times of low demand by acting as a natural "sink" for renewable energy once it becomes cost-competitive for peaking applications. This could therefore result in a positive feedback loop in which renewable energy sources raise the market potential for storage, which raises the chances for renewable energy sources to avoid production of very low-value energy and curtailment. As variable renewable energy sources become more widely used, there will be a greater demand for additional services that battery energy storage can offer.

For instance, growing renewable energy will affect the following grid services, most of which energy storage (along with other technologies) can mitigate:

- Regulation reserves will need to increase to balance the variability inherent in wind and solar resources.

- Fast-frequency response may be required to replace system inertia as conventional synchronous generators are utilized less or retire.
- Voltage support (volt/var support) needs to be managed across the system, especially as renewable resources are added to more remote locations.
- Energy shifting and opportunities for price arbitrage will increase to balance periods of high renewable energy penetration (sunny and windy days) and periods of low penetration (cloudy and calm days). This is especially true for solar-centric renewable energy grids, which are characterized by a diurnal pattern and shorter periods of renewable energy availability.

Battery energy storage is only one form of power system flexibility available to renewable energy generators, utilities, and grid operators. It should be noted the first three services mentioned above can also be provided by state-of-the-art wind and solar power plants themselves. In addition, while increasing renewable penetration will increase the number of required ancillary services, such as regulation, that increase is small relative to the amount of wind and solar generation added to the system.

For example, recent renewable integration studies suggest that when adding significant levels of wind capacity to a grid, only 0.3 to 1MW of reserves are needed to balance variability from 100 MW of additional wind. Therefore, market saturation for ancillary services happens quickly, and the market size of regulation and other ancillary services may not grow to a level required to scale energy storage applications.

Eventually, the economics of battery energy storage will likely accelerate deployment across the grid; however, other technologies are providing this flexibility as well. Demand response, advanced inverter functionality, flexible AC transmission systems (FACTS), and improved flexibility from the thermal generating fleet and renewable energy plants will also provide some of the services listed above. As a result, it is unclear whether or not the economic value of providing ancillary services will increase along with the need for new services (Table 8.5).

Table 8.5. *Main Advantages and Disadvantages of Battery Energy Storage Systems (BESSs) (As the Table Has Been Reproduced from the Cited Source, Please Note That the References in Brackets Have Been Cited in the Paper)*

Technologies	Advantages	Disadvantages
Pb-acid	Mature technology low cost tolerance against misuse low maintenance requirements large storage capacity spill-proof fast response time low daily self-discharge rate	Bulky (low energy density and specific energy increased cost under low temperature operation reduced lifespan at prolonged discharged state low cycling capability low DOD hazardous for the environment
Li-ion	High energy density, high power density, high efficiency, long life cycle, low discharge rate, no-memory effect, stable discharge voltage wide operating temperature, high cycle efficiency, packaging flexibility, recyclable lithium oxides and salts, high specific energy, rapidly decreasing costs, excellent charge retention, high cell voltages, very good performance at low temperatures high DOD	Increased cost due to protection circuits and packaging, availability, safety and environmental issues, temperature-dependent life cycle
VRF	Long life cycle flexible design of power and energy capacity, low standby losses, simple cell management, high energy efficiency, low maintenance cost, large storage capacity, high power output and energy conversion rate, suitability for large-scale energy storage, low self-discharge, durable performance, overcharge tolerance, high DOD	High cost large layout high capital and running costs unsuitable for small-scale storage applications low energy density

Tech-nolo-gies	**Advantages**	**Disadvantages**
NaNi-Cl_{22}	High energy density long life cycle maintenance-free long discharge time fast response high efficiency	High cost high self-discharge heat required to keep the molten state temperature

Source: Mexis I, Todeschini G. Battery Energy Storage Systems in the United Kingdom: A Review of Current State-of-the-Art and Future Applications. Energies. 2020; 13(14):3616. https://doi.org/10.3390/en13143616.

8.2.5. The Future Ahead, Obstacles, and Opportunities

8.2.5.1. The Role of Policy, Regulations, and Markets in the Energy Storage Industry

Battery energy storage will not be quickly adopted and deployed if technological advancements alone continue. The incapacity of energy storage to profit from every value stream it can offer is one of the main challenges it faces. Although it is generally acknowledged that energy storage offers a number of benefits that could be helpful to the electrical grid, these benefits are frequently hard to measure and fairly price. Energy storage mandates have generally been the primary means of driving recent policy, which has primarily come from state levels. As of 2016, laws requiring or offering subsidies for battery energy storage were in place in California, Connecticut, Massachusetts, Nevada, New York, and Oregon. The California Public Utilities Commission, in particular, mandated that by 2020, the state's regulated utilities acquire a minimum of 1325 gigawatt-hours of energy storage. A minimum quantity of battery energy storage capacity is required by legislation that is either in place or being proposed in other states. These regulations may be an ineffective means of obtaining all the benefits that energy storage offers, even though they will be successful in encouraging pilot projects.

To better support battery energy storage technology deployment, policymakers, regulators, and market operators should focus on three broad topics that can have a significant impact:

1. Tailor regulations and markets for desired results.
2. Make policies, regulations, and market rules technology-agnostic.
3. Pursue a diverse mix of hybrid storage projects.

8.2.5.2. Tailor Regulations and Markets for Desired Results

Acquiring storage's valuable services, such as fast frequency response, regulation, ramp rate support, peak load reduction, energy arbitrage, voltage support, and postponing transmission and distribution upgrades, is the aim of battery energy storage policy. The utility or grid operator will need to properly control and optimize the energy storage asset for minimum capacity targets (RPS) for energy storage to deliver those services. Markets and initiatives for grid services will guarantee a more steady development of energy storage projects, while minimum installed capacity requirements will aid in the development of pilot projects and promote industry learning. It is also important for regulators, policymakers, and market participants to concentrate on giving desired services a monetary value. Sometimes grid services are valuable to electricity consumers and grid operators, but there is no market mechanism or financial incentive in place for them. For instance, a lot of regulated utilities and deregulated power markets lack a particular ramp rate product. Together with other technologies, this would give battery energy storage another source of income if it were implemented. By placing a higher value on battery energy storage services than on minimum capacity targets, policies, and incentives can be made more flexible and adaptable to changes in technology and the mix of resources in the power system. For instance, there is enough system inertia in many of the large, interconnected power grids of today, so fast frequency response is not very valuable economically. But in a future resource mix where electronic generators with higher power (i.e., wind, and solar), there might be a greater requirement for quick frequency response. Establishing a framework for the current and future monetization of this service, even at a discounted rate, will incentivize battery energy storage providers to offer it.

8.2.5.3. Make Policies, Regulations, and Market Rules Technology Agnostic

Energy storage technology is developing quickly, with various forms of storage currently available such as compressed air, pumped, and battery energy storage. Furthermore, controllable loads can function as virtual storage devices without requiring significant upfront costs. Therefore, it is important to avoid favoring one technology over another when creating policies and regulations. Instead, they should be technology neutral. Technical specifications are always crucial for defining the needs of the power system, but they should be drafted with the system's physical requirements—such as response time, duration, and other factors—in mind) and not on the features of the technology that is currently accessible.

8.2.5.4. Pursue a Diverse Mix of Storage Projects and Unique Applications

Storage developers will have to explore innovative applications for additional deployment until the costs and market mechanisms of battery storage are aligned. The hybridization of battery energy storage with other generation technologies is a recent trend in the power industry. For instance, solar+storage plants, which combine battery energy storage and solar generating, are being developed. These plants frequently share infrastructure within the DC subsystem or even at the point of interconnection. Costs associated with transmission interconnection, project design and engineering, construction, and development (environmental, engineering, etc.) can all be reduced by doing this), and land cost, providing additional chances to group projects and realize economies of scale. Hybrid projects offer a streamlined and simplified acquisition process for utility off-takers, enabling them to execute a single PPA for renewable generation and storage.

The U.S. federal Investment Tax Credit, which offers a tax incentive to offset the capital cost of solar installations, is the primary reason behind the hybridization of battery energy storage with solar PV projects in the US. There are other hybrid plant configurations as well, such as energy storage combined with conventional gas turbines or utility-scale wind plants. When energy storage is combined with a traditional gas turbine, for instance, the unit becomes more flexible, has instantaneous startup capabilities, and can take part in more ancillary service markets. Policies that support hybrid projects in the future should be technology agnostic and enable storage to be integrated with a wide range of conventional and renewable generation resources.

Hybrid configurations may result in an inefficient amount of storage on the system, even though they let plant developers split the cost of storage deployment with a generating unit. Storage is generally better understood as a system asset than as a generator asset on its own. For example, to mitigate the variability brought about by a new wind or solar plant, certain utilities and grid operators need storage. Managing the variability at each wind and solar plant on the system will result in a large overbuild and the loss of the advantages of geographic diversity. On a cloudy day, the output of a single solar PV plant will be highly variable, but the output of multiple plants spread throughout the area will be noticeably smoother. To put it briefly, when a single plant is balanced using storage, power may be injected at one grid point and removed at another plant location by a nearby storage system. As a result, storage ought to be thought of as a system asset rather than as a single plant that regulates frequency, voltage, ramp rate, and energy shifting for the balancing area.

8.2.6. The Future Ahead for Energy Storage and Renewable Integration

Global power grids will continue to use more and more battery energy storage. As battery energy storage costs continue to drop, industry learning and acceptance solidifies through the development of pilot projects, and the need for ancillary services grows as the resource mix continues to shift from a synchronous, fossil-fuel-based grid to a future renewable-centric grid, these trends are likely to accelerate. The industry's focus on battery energy storage has recently increased, but this shouldn't take away from resource planning that is fundamentally sound and based on the financial and physical needs of ratepayers as well as the needs of the grid. Though there are numerous other options, battery energy storage is one type of grid flexibility that will allow for the modernization of our electrical grid and the increased use of renewable energy. Prioritizing the "low hanging fruit" for system flexibility over relatively capital-intensive battery energy storage projects is a good strategy for grid planners and operators. The following are just a few examples of these additional forms of flexibility.

1. Conventional thermal unit flexibility: improved cycling (on/off) performance at lower cost, reduced minimum stable operating levels (P-Min), and faster ramp rates.
2. Hydro unit flexibility and coordination with wind and solar output levels.
3. Advanced inverter functionality from wind and solar resources, both at the utility-scale and distribution levels to include bi-directional control and communication, low-voltage ride-thru, volt-var support, frequency ride-thru, fast-frequency response, and regulation.
4. Demand response, including energy shifting and for essential reliability services like fast frequency response, regulation, and replacement reserves.
5. Increased demand-side price responsiveness, via time-of-use or real-time pricing.
6. Increased cooperation and coordination amongst neighboring balancing areas and across interconnections via HVDC lines.

One of the many important tools and technologies that will help modernize the grid and encourage more renewable energy penetration is battery energy storage. Large-scale energy shifting between time periods is not necessary for the majority of bulk power systems in use today because of excess renewable generation or notable differences in on-peak and off-peak electricity prices. Although most places do not currently have a need for multi-hour (4–6 hour) duration battery systems, long-term trends in the power industry will favor their deployment in the future. Thus, in areas where the value of grid reliability services is greatest, such as island and remote grids, municipal, and cooperative electricity providers, and regions with a high penetration of renewable energy, near-term battery energy storage deployments should continue to prioritize delivering these services. In the end, battery energy storage will become more important in the markets for energy, capacity, and ancillary services as the power system gets more modernized and renewable energy sources become more prevalent (Akhil et al., 2013; Bhatnagar et al., 2013).

8.2.7. Impact on Carbon Footprint

Achieving climate goals and lowering the carbon footprint of power systems depend heavily on the integration and storage of renewable energy. We may greatly reduce our dependency on fossil fuels and subsequently reduce greenhouse gas emissions by integrating renewable energy sources like wind, solar, and hydropower into the grids that currently operate. We can also do this by utilizing a variety of storage technologies.

8.2.7.1. Renewable Energy Integration

The need for energy from fossil fuel-based plants, which are significant contributors to carbon dioxide (CO_2) emissions, is replaced when renewable energy sources are integrated into power grids. For example, solar, and wind energy generate electricity without releasing greenhouse gases into the atmosphere. These sources contribute to lowering the total carbon intensity of electricity generation as they are used more frequently. The switch to renewable energy sources lessens reliance on carbon-intensive energy sources like coal, oil, and natural gas, which are major contributors to global warming.

8.2.7.2. Energy Storage Technologies

Efficient energy storage systems are necessary to control the erratic and intermittent nature of renewable energy sources. Systems for storing and releasing excess energy produced during high production periods include lithium-ion batteries, pumped hydro storage, and flow batteries. This feature guarantees the continuous use of renewable energy, even in the absence of power generation from the sources. Storage systems lessen the need to rely on backup fossil fuel plants, which would otherwise step in during times of low renewable generation, by enabling a steady supply of renewable energy.

8.2.7.3. Carbon Footprint Reduction

When renewable energy and storage technologies are combined, the electricity sector's carbon footprint is significantly reduced. According to a study conducted by the International Renewable Energy Agency (IRENA), switching to a high-renewable energy mix has the potential to reduce global CO_2 emissions by up to 70% by the year 2050. Higher renewable energy penetration is made possible by energy storage, which also improves the overall power system's efficiency and lowers overall emissions from energy generation and transmission.

8.2.7.4. Economic and Systemic Benefits

Beyond the direct emissions reductions, integrating renewable energy and storage can lead to broader economic and environmental benefits. It supports the transition to a low-carbon economy, creates jobs in the renewable sector, and reduces air pollution, which has additional health and environmental benefits.

The integration and storage of renewable energy are critical for reducing the carbon footprint of energy systems. They enable a cleaner, more stable, and reliable power supply while reducing dependency on fossil fuels and lowering greenhouse gas emissions, thereby contributing significantly to global climate change mitigation efforts (Akhil et al., 2013; Bhatnagar et al., 2013).

8.3. THE FUTURE OF ENERGY STORAGE

One important component of mitigating climate change is energy storage. Just 3% of the world's power capacity is being stored. Energy storage capacity needs to triple by 2050 to keep global warming to less than 2°C. To accomplish this, we must consider creative ways to expedite the procedure. Future Energy Storage" examines the potential benefits of energy storage as well as the challenges facing its adoption. Could energy-storage-as-a-service be one of the answers? We can use energy more efficiently and reduce carbon emissions by using energy storage.

Energy storage capacity would need to rise from 140 GW in 2014 to 450 GW in 2050 to keep global warming to less than 2°C. Just 3–4% of the electricity produced by utilities worldwide is stored at this time.

A revolution in energy storage is about to happen, but there's a chance that a number of roadblocks will prevent it from happening. Common misconceptions regarding the costs, safety, and commercial availability of storage technologies may be the most difficult obstacles to overcome.

Certain social groups in our societies are more open to embracing new technologies than others. In the 1960s, Everett Rogers popularized the adoption curve and the diffusion of innovations theory. As per this theory, a technology's ability to spread beyond its early adopters reaches its tipping point when it reaches a market share of approximately 15 to 18%. Experts concur that one of the most important components of the plan to combat climate change is energy storage. Still, it appears that the new energy storage technologies are not quite at the tipping point. The reasons behind this and potential solutions are examined in the paper "Beyond the Tipping Point: Future Energy Storage."

8.3.1. Efficient Energy Storage to Reduce Emissions

Energy storage can reduce emissions from electricity, heating, and cooling requirements, hence contributing to the global fight against climate change. Through the integration of locally produced energy, energy storage at the community level can involve more adaptable and resilient energy systems with higher levels of energy security. Energy storage offers citizens the advantage of better control over energy costs and origin. With ongoing service innovation, energy storage can create new business opportunities for the private sector. These can include market players growing into the installation of storage systems, charging stations, and virtual power plants, as well as businesses offering energy storage as a service.

If we want more people and organizations to use energy storage, we need to make it easy and cost-efficient. Energy-storage-as-a-service is a very promising concept that could achieve that.

In the energy-storage-as-a-service model, energy becomes available to the customer as a service, in the same way as subscriptions for food, accessories, films or music. Energy-storage-as-a-service can curb the barrier of high upfront investment costs, help aggregate multiple storage revenue sources, and give the public the opportunity to test a solution without committing to a long-term investment.

Why is effective energy storage important? One reason is that energy storage can balance the supply and demand of renewable energy. Wind and solar power are both variable energy sources; simply put, energy storage can save the day when the sun doesn't shine or the wind doesn't blow, by feeding stored energy from sunnier or windier days to the grid.

Another example relates to the excess heat of our homes. Storage can be used for buildings that have both heating and cooling demands – such as offices, shopping centers, hospitals, and airports. Storing your own waste heat and cooling for later use in the season when it is needed can save primary energy and as well as money.

8.3.2. Energy Storage: the Key to a Decarbonized Future

Efficient energy storage is a fundamental pillar of the energy transition: allowing flexible renewable energy production and guaranteeing its integration into the grid. Find out which storage systems are the most efficient and which ones promise to drive the much-needed transition towards a decarbonized electricity system.

Energy storage not only integrates renewable but also saves the electrical grid energy. It is simple to produce, move, and transform electricity. But, there hasn't been a practical, simple, or economical way to store it up until now. Accordingly, to integrate renewable energies and prevent drops in clean energy during supply troughs, as well as to increase the efficiency and security of the electrical grid, it is necessary to continuously generate electricity based on demand.

Improved electrical energy storage is essential to support renewable energy sources like wind and solar energy, which are replacing fossil fuels in the world. This will allow electrical grids to remain balanced and maximize the amount of green megawatts generated.

8.3.2.1. Main Energy Storage Systems

Electricity cannot be stored as such and therefore it needs to be transformed into other types of energy, such as mechanical or chemical. Storage systems can add value at every stage of the supply chain. Depending on their capacity, energy storage systems are divided into: large-scale storage, which is used in places where GW scale is required; storage in the grid and in power generation assets, where the MW scale is used and, finally, storage at the end-user level, which applies to the residential level and works with kW.

Currently, the following list shows the ways to accumulate energy and the main technologies that allow efficient energy transformation and storage:

1. **Hydroelectric Pumping:** The most efficient large-scale storage system in operation. This is a cost-effective and proven technology that provides stability to the electrical system and can generate significant levels of clean energy with rapid response times.
2. **Compressed Air:** These facilities have a reversible motor that, during periods of excess energy, ambient air is stored at high pressures in underground cubicles. It is a mechanical storage system comparable in capacity to hydroelectric pumping.
3. **Thermal Storage:** It consists in accumulating energy in materials that allow it to be retained and released in a controlled manner,

through methods ranging from cooling through ice accumulation to exposure to extremely high temperatures.

4. **Supercapacitor:** This is a device capable of storing large amounts of electrical energy in the form of electrostatic charges, meaning there are no chemical reactions. Super capacitors can be charged or discharged in a matter of seconds, thus being ideal for responding to energy peaks or brief supply interruptions.
5. **Flywheels:** This is a mechanical storage system consisting of a metal disc that starts to spin when a torque is applied to it, then, with the action of braking the wheel by applying a firm braking torque, electrical energy is conserved in kinetic form.
6. **Hydrogen Fuel Cells:** This is a type of continuous chemical storage. It differs from batteries in that it is supplied continuously with hydrogen from the outside allowing its constant use. There are other types of fuel cells, but hydrogen is the most commonly-used fuel.

8.3.3. Lithium Batteries: The Future of Storage

Lithium-ion batteries have been viewed by the renewable energy industry as the answer to its primary issue in recent years, which is the storage of generated energy. Lithium, one of the smallest elements in the periodic table, has a high potential for electrochemistry and has a large energy storage capacity. Lithium batteries have the desired low weight and high efficiency, but their high cost has kept them from taking the lead as the go-to storage technology for renewable energy.

That said, things appear to be shifting in this regard. A recent study conducted by Bloomberg NEF (BNEF) predicts that lithium-ion battery costs will drop dramatically in the upcoming years, even more than the 85% decrease that took place between 2010 and 2018. In particular, BNEF projects that by 2030, the cost of lithium-ion batteries per kW/h will have dropped by 50% due to the explosive growth of two distinct markets: electric vehicles and stationary storage.

This will lead to worldwide energy storage facilities multiplying exponentially, from modest 9GW/17GWh implemented from 2018 to 1,095 GW/2,850 GWh by 2040. This dramatic increase will require an estimated investment of US$662 billion (Figure 8.2).

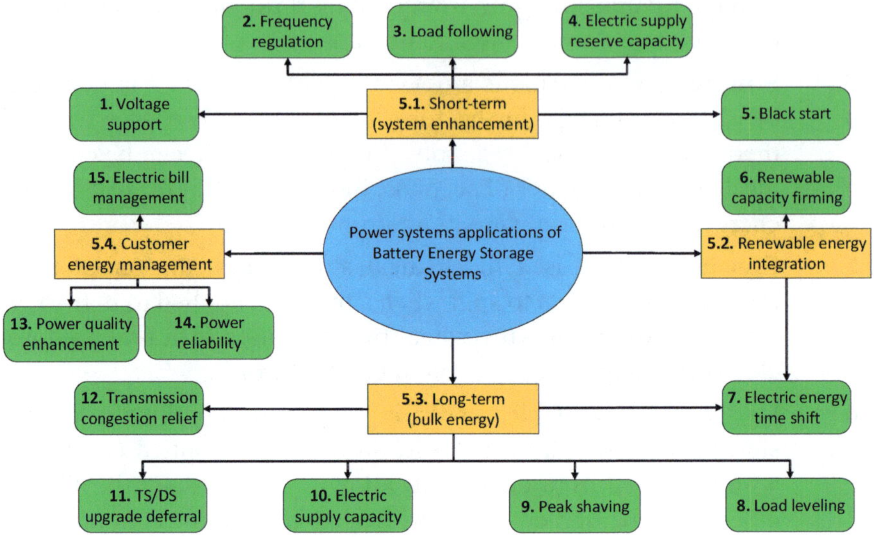

Figure 8.2. *Classification of the ancillary services provided by battery energy storage systems (BESSs) connected to the transmission and distribution system in the UK.*

Source: Mexis I, Todeschini G. Battery Energy Storage Systems in the United Kingdom: A Review of Current State-of-the-Art and Future Applications. Energies. 2020; 13(14):3616. https://doi.org/10.3390/en13143616.

According to BNEF forecasts, the total demand for batteries in the sectors of stationary storage and electric vehicles will be 4,584 GWh by 2040, which, in addition to supporting the energy transition, will be a great opportunity for battery manufacturers and extraction companies of the components required, such as lithium, cobalt, and nickel.

8.3.4. Energy Storage Myths in the Way of Development

There are several myths standing in the way of energy storage reaching its full potential. Through debunking these myths, we can make energy storage an integral part of the future energy system. One energy storage myth is that we have to wait for the perfect solution.

8.3.4.1. Battery Storage

A battery that offers unparalleled performance, a high energy density, affordable prices, minimal environmental impact, and an extended lifespan. If fortune favors us, these batteries may appear on store shelves at some point in the future—just not as soon as the media suggests. It will take at least 10 years for the battery that is currently being developed in a lab to be made commercially available. After extensive testing to guarantee operational safety, achieving economies of scale presents a challenge.

8.3.4.2. Geothermal Storage

The myth surrounding geothermal energy is another. In Iceland, people typically only connect geothermal energy with geysers. But geothermal energy comes in two main forms. Shallow geothermal energy is solar energy that has been stored in the ground, whereas deep geothermal energy originates from the Earth's core. While shallow geothermal energy, which is typically found between 0 and 300 meters below the surface, is readily available everywhere, deep geothermal energy utilization typically requires drilling very deep boreholes close to natural cracks.

The yearly average air temperature and the ground temperature are roughly equal throughout the year. As a result, the ground temperature in Berlin is 9°C, in Stockholm it is 6°C, and in Helsinki it is 5°C. These are not high enough temperatures to be utilized for building heating. But since it's significantly warmer than the outside air in the winter, this can be used as preheating.

8.3.4.3. Hydrogen Storage

The idea that hydrogen is hazardous is another widespread misperception. But when it comes to safety protocols and regulations, hydrogen is just like any other fuel. In contrast to gasoline, hydrogen leaks quickly spread throughout the atmosphere. At room temperature, hydrogen is not reactive, and when it burns, water vapor is produced. Hydrogen has not yet become widely used, despite its versatility and ability to store renewable energy giving it promising potential. At the moment, hydrogen can only be used for small-scale storage applications because it is primarily produced using non-emission-free processes.

The public and private sectors must work together to build the infrastructure necessary for refueling for citizens to embrace hydrogen to a greater extent. An annual amount of CO_2 emissions equal to the combined emissions of Indonesia and the United Kingdom is caused by hydrogen production. As renewable energy sources are deployed more widely and our overall electricity consumption rises, hydrogen's current supporting role should give way to a leading one. Both small- and large-scale hydrogen storage are possible thanks to this technology, which can store energy locally and distribute it via global supply chains. To make sure that demand is appropriately satisfied, the required fueling infrastructure must be simultaneously rolled out.

8.3.5. Role in an Equitable Clean Energy Future

Similar to how food could be refrigerated for several days or weeks before being consumed or thrown away, energy storage enables people and communities to have access to electricity when they most need it—during blackouts or in the absence of sunlight, for example. Storage has the potential to lessen the need for electricity produced by dirty, inefficient plants, many of which are found in underprivileged and marginalized areas. Additionally, storage can help balance demand, protecting electricity consumers from sudden increases in price.

In the intricate system that is the electrical grid, supply, and demand for power must always be equal. Throughout history, supply has adapted to meet demand, which has resulted from both predictable changes like equipment overloads, wildfires, storms, and other extreme weather events, as well as daily patterns of human activity. These days, we also consider electricity demand flexibility to maximize the use of renewable energy sources, including how and when we heat and cool our homes and charge electric cars. In this delicate balancing act, energy storage contributes significantly to the development of a more adaptable and dependable grid system. For instance, excess electricity generation can be used to charge storage devices when there is a surplus of supply compared to demand, such as at night when steadily running power plants provide electricity or during the middle of the day when the sun is at its brightest. Storage facilities, including those found in people's homes, can release their stored energy into the grid when demand exceeds supply.

For many years, hydroelectric dams have been utilized as storage devices by pumping water back behind them. This allows the electricity generated by the turbine to be released later, when needed, and absorbs excess generation from the grid. Large battery banks made of lithium-ion batteries are now more frequently found in residences, neighborhoods, and utility buildings.

8.3.6. What Is Energy Storage and How Does It Work?

To put it simply, energy storage is the capacity to store energy for use at a later date. Energy-saving options for storage devices are numerous (e.g., chemical, kinetic, or thermal) and transform them back into electrical power or other useful forms of energy.

Numerous storage technologies are currently in use, despite the fact that pumped hydro accounts for almost all energy storage capacity today and that battery system deployment is accelerating quickly. The facilities known as pumped hydroelectric storage are situated along the transmission lines of the grid. They store excess electricity and can react to the grid's demands in less than 10 minutes. Pumped hydroelectric storage converts the kinetic energy of falling water into electrical power. When supply outpaces demand, the systems, which have two reservoirs at different elevations, store energy by pumping water into the upper reservoir. When supply cannot keep up with demand, water is released into the lower reservoir and runs through turbines to create energy.

While some new projects are being planned, the majority of pumped hydro systems were constructed in the 1970s to go along with the new generation of nuclear power plants. Nuclear power plants generate energy continuously throughout the day because they are not meant to ramp up or down. Pumped hydro facilities store excess electricity for use during peak demand hours, when demand for electricity is lower. Pumped hydro plants have shown to be useful in promptly adapting to slight variations in supply or demand.

8.3.6.1. Batteries

Through electro-chemical processes, batteries store energy, transforming it into chemical energy and back again when needed. Lead-acid, metal-air, lithium-ion, and sodium-sulfur batteries are among the varieties. Because of their high energy density, high power, and high efficiency, lithium-ion batteries—like the ones found in laptops and cell phones—are one of the energy storage technologies that is expanding the fastest. At the moment, lithium-ion battery applications at the utility scale are limited to four hours of power.

Depending on the type, dimensions, functionality, and needs of the house, residential storage may endure longer. Batteries can be placed in the distribution system, at the generator, or alongside transmission lines. Additionally, they have a wide range of applications, including in homes, businesses, and electric cars.

Technological developments in lithium-ion batteries have been primarily attributed to the growing electric vehicle (EV) market. Lithium-ion batteries require a number of uncommon but crucial materials. These materials raise global environmental justice issues, like those involving cobalt. This emphasizes the necessity of transparent supply chains for battery production, as well as strict environmental and labor laws governing all vital material extraction procedures. Equally important is ongoing research and development to find alternative materials or technologies (like zinc-air batteries) that could wean the battery industry off of its reliance on mining for essential minerals, particularly in regions with lax labor and environmental regulations or a history of human rights abuses.

8.3.6.2. Thermal Storage

A system known as concentrated solar power (CSP) uses mirrors or lenses to gather solar energy. The concentrated sunlight is then used to heat a fluid, which powers a turbine to produce electricity. Thermal storage is the practice of storing heat for later use as opposed to using it instantly to create electricity. The hot fluid is kept at a high temperature in sizable tanks until it is needed. It can be water, molten salts, or other molten materials. Solar energy can be collected and concentrated using a variety of designs.

The majority of CSP facilities in the United States—including the 399-MW Ivanpah Solar Power Facility, one of the biggest in the world—are found in the southwest desert region. Systems that counteract the need for electricity rather than using it to create it directly are also referred to as thermal storage systems. One method is to use off-peak electricity and air conditioning to freeze water at night. The ice is melted and cool air is blown over the air conditioning condenser coils during the day when there is a high demand for cooling, which lowers the amount of electricity required to keep the building cool. Numerous commercial buildings, such as the University of Arizona and the state government buildings on the North Carolina capitol campus, use these kinds of systems.

8.3.6.3. Compressed Air

A device known as compressed air energy storage, or CAES, compresses air using extra electricity and stores it, typically in an underground cavern. The compressed air is released and used to power a turbine to generate electricity. Typically, a gas turbine's compressor is powered by compressed air in a CAES design, saving roughly two thirds of the energy required to run the turbine itself. This results in less natural gas being used, and depending on the design, it can reduce carbon dioxide emissions by 40 to 60%. CAES systems are long-lasting, have a high storage capacity, and a high power rating. However, there aren't many places that are suitable for CAES plants because they need an underground reservoir.

8.3.6.4. Flywheels

Electricity is transformed into rotational kinetic energy stored in a spinning mass by flywheel energy storage systems. To lessen drag, the flywheel has a big rotor inside a vacuum inside a cylinder. A motor powered by electricity accelerates the rotor to 60,000 rpm, one of the highest speeds possible. The motor functions as a generator to release the stored energy by turning the kinetic energy back into electrical energy. Flywheels usually require little maintenance and have a long lifespan. The gadgets are also very efficient and respond quickly. Flywheels can be positioned close to the consumers and store electricity for distribution because they can be positioned almost anywhere. A "flywheel farm" is a collection of connected flywheels that can create a storage facility with a capacity of megawatts, although a single flywheel device typically has a capacity of kilowatts or less. Currently, Beacon Power runs the two biggest flywheel short-term energy storage facilities in the country, one each in Pennsylvania and New York. With a 20 MW operating capacity, each plant is mainly utilized for frequency regulation, which helps to balance variations in the supply and demand of electricity.

8.3.6.5. Hydrogen

When water is electrolyzed (broken down into hydrogen and oxygen) using renewable electricity, hydrogen can be stored as a clean energy source. Large amounts of hydrogen can be kept in subterranean caverns, while smaller amounts can be kept in storage tanks. Later on, stored hydrogen can be put to many different uses, such as chemical feedstocks or maritime transportation. It can be converted back into electricity using combustion turbines or fuel cells, the latter of which only produce water as a byproduct while burning hydrogen can release harmful Nox emissions (Akhil et al., 2013; Bhatnagar et al., 2013).

REFERENCES

1. Akhil, A., Huff, G., Currier, A., Kaun, B., Rastler, D., Chen, S. B., Cotter, A., et al. (2013). *Electricity Storage Handbook* (SAND2013-5131). U.S. Department of Energy (DOE), Electric Power Research Institute (EPRI), National Rural Electric Cooperative Association (NRECA).
2. Bhatnagar, D., Currier, A., Hernandez, J., Ma, O., & Brendan, K. (2013). *Market and Policy Barriers to Energy Storage Deployment* (SAND2013-7606). Sandia National Laboratories, Albuquerque, NM.
3. California Public Utilities Commission (CPUC). (2018). *Decision on Multiple-Use Application Issues* (Rulemaking 15-03-011).
4. Cochran, J., Miller, M., Zinaman, O., Milligan, M., Arent, D., Palmintier, B., O'Malley, M., et al. (2017). *Flexibility in 21st Century Power Systems* (NREL/TP-6A20-61721). National Renewable Energy Laboratory (NREL).
5. Curry, C. (2017). *Lithium-Ion Battery Costs and Market* [Market report]. Bloomberg New Energy Finance.
6. Denholm, P. (2018). *Batteries and Storage: Truly a Game Changer?* Paper presented at the JISEA 2018 Annual Meeting, Golden, CO.
7. Denholm, P., & Margolis, R. (2018). *The Potential for Energy Storage to Provide Peaking Capacity in California Under Increased Penetration of Solar Photovoltaics* (NREL/TP-6A20-70905). National Renewable Energy Laboratory (NREL).
8. Lew, D., Brinkman, G., Ibanez, E., Florita, A., Heaney, M., Hodge, B. M., Hummon, M., et al. (2013). *The Western Wind and Solar Integration Study Phase 2* (NREL/TP-5500-55588). National Renewable Energy Laboratory (NREL).
9. Mills, A., & Wiser, R. (2012). *Changes in the Economic Value of Variable Generation at High Penetration Levels: A Pilot Case Study of California* (LBNL-5445E). Lawrence Berkeley National Laboratory.
10. Palchak, D., Cochran, J., Ehlen, A., McBennett, B., Milligan, M., Chernyakhovskiy, I., Deshmukh, R., et al. (2017). *Pathways to Integrate 175 Gigawatts of Renewable Energy into India's Electric Grid, Vol. I—National Study*. National Renewable Energy Laboratory (NREL).

Chapter 9

The Safety of Energy Sources and Future Trends and Innovations

Contents

INTRODUCTION

One of the most important factors supporting the sustainability and dependability of our global energy infrastructure is the safety of energy sources. Conventional sources, like fossil fuels, have long been linked to a number of safety issues. The urgent need to switch to greener options is highlighted by oil spills, air pollution, and its role in climate change. Although they present promising solutions, renewable energy sources like solar and wind power also have safety concerns. For example, photovoltaic panels in solar energy systems contain potentially hazardous materials that need to be handled and disposed of carefully to avoid harm to the environment. Despite being environmentally benign, wind turbines need careful safety precautions to be taken during construction and operation. These buildings' immense size and height necessitate stringent regulations to protect the workers' and the public's safety. Furthermore, worries about how wind turbines affect wildlife highlight the significance of careful risk assessments and mitigation plans.

Despite being a powerful energy source, nuclear energy raises a number of safety issues. Comprehensive safety measures are necessary due to the possibility of catastrophic accidents, the management and disposal of radioactive waste, and other factors. Strict legal frameworks and cutting-edge innovations, like reactor designs with failsafes, are necessary to reduce these risks and increase public trust in nuclear power. Finding a careful balance between maximizing the potential of various energy sources and guaranteeing their safety becomes crucial as the world struggles to address climate change and meet rising energy demands. Technological advancements that improve energy production, transmission, and consumption safety include smart grid systems and sophisticated monitoring technologies.

Furthermore, it is essential to promote a safety culture in the energy sector. This calls for strict adherence to safety procedures, ongoing training, and a dedication to using lessons learned from previous incidents to stop them from happening again. Public engagement and awareness are equally important because educated citizens can support and help to implement strong safety regulations.

Energy source safety is a complex issue that calls for an all-encompassing and proactive strategy. Energy production must put community well-being, environmental preservation, and human safety first, regardless of whether it comes from new or conventional fossil fuels. We can negotiate the complexity of our energy landscape and create a sustainable, safe, and resilient future by continuing research, innovating, and uniting behind strict safety regulations (Buonocore et al., 2015; Cacciapaglia et al., 2018).

9.1. RENEWABLE ENERGY SAFETY

Although renewable energy sources—like solar, wind, hydro, and geothermal power—are praised for their positive effects on the environment, it is imperative that their development, use, and upkeep be done safely. Within the solar energy industry, handling potentially hazardous materials during the production and installation of photovoltaic panels calls for stringent safety procedures to safeguard both the environment and laborers. Towering turbines associated with wind energy necessitate extensive safety precautions during construction and ongoing maintenance. It is crucial to provide employees working at heights with rigorous training, and to implement secure access procedures. Although hydroelectric power is a clean energy source, dam safety must be carefully considered to avoid catastrophic failures and safeguard communities downstream. In a similar vein, the extraction of geothermal energy necessitates controlling hot fluids and possible underground risks (Figure 9.1).

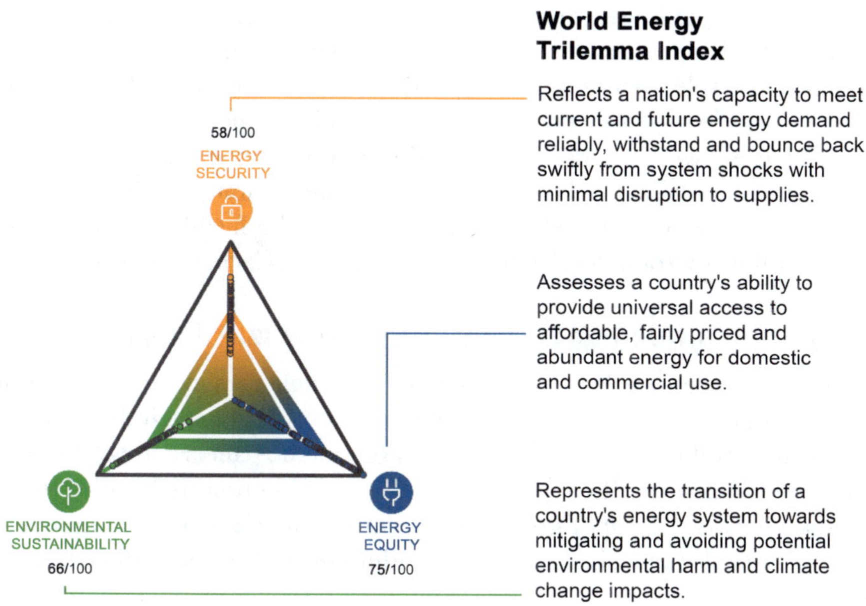

Figure 9.1. *Clean energy supply.*

Source: Elkhatat A, Al-Muhtaseb S. Climate Change and Energy Security: A Comparative Analysis of the Role of Energy Policies in Advancing Environmental Sustainability. Energies. 2024; 17(13):3179. https://doi.org/10.3390/en17133179.

The production and disposal of parts, like batteries in storage systems, present environmental and health risks that require careful consideration in the larger framework of renewable energy. When the renewable energy industry grows, maintaining a strong safety culture becomes crucial. Strict training regimens, thorough risk analyses, and the creation of uniform safety procedures all help to reduce mishaps and guarantee employee welfare. It is essential to conduct routine maintenance and monitoring to spot possible problems early on and take appropriate action. Furthermore, to maintain grid stability and avoid mishaps, the integration of renewable energy sources into current power grids necessitates meticulous planning.

Adopting cutting-edge safety features like sophisticated monitoring systems and fail-safe mechanisms becomes essential as technologies advance. A further factor in encouraging the responsible development of renewable energy projects is international collaboration in the sharing of best practices and the establishment of safety standards.

An essential component of the safety of renewable energy is public engagement and awareness. Cooperation and understanding are promoted when communities are made aware of the possible risks connected to renewable energy projects and the safeguards put in place to reduce those risks. In conclusion, although renewable energy sources greatly advance a sustainable future, it is critical to prioritize safety throughout these technologies' whole life cycles. Regulatory frameworks, a robust safety culture, and technological advancements work together to create a holistic approach that guarantees the benefits of renewable energy are realized without sacrificing the health of communities, workers, or the environment (Buonocore et al., 2015; Cacciapaglia et al., 2018).

9.1.1. Photovoltaic Panel Recycling and Hazardous Materials

The increasing use of solar photovoltaic (PV) panels as a clean energy source has drawn attention to the environmental issues surrounding their disposal at the end of their useful life. Materials used in photovoltaic panels can potentially be hazardous to human health and the environment if improperly handled. One of the main issues is the use of potentially harmful materials in the production of some kinds of solar cells, including lead, cadmium, and occasionally hexavalent chromium.

Cadmium telluride (CdTe) and copper indium gallium selenide (CIGS) are thin-film technologies that use cadmium and other potentially hazardous substances. Meanwhile, crystalline silicon solar panels, the most common type, typically contain small amounts of lead. The hazardous nature of these materials raises concerns about the potential for environmental contamination during the disposal or recycling process.

To address these concerns, proper recycling and disposal practices for end-of-life photovoltaic panels are essential. Recycling offers an opportunity to recover valuable materials from decommissioned panels while minimizing the environmental impact of hazardous substances. Various recycling technologies and processes have been developed to extract and treat these materials safely.

The recycling process typically involves the following steps:

1. **Collection and Transportation:** Collecting decommissioned solar panels and transporting them to recycling facilities.
2. **Dismantling:** Removing the panels from their mounts and separating the various components.
3. **Material Recovery:** Extracting valuable materials, such as silicon, silver, and aluminum, for reuse.
4. **Hazardous Material Treatment:** Managing hazardous materials through specialized treatment processes to prevent environmental contamination.
5. **Safe Disposal:** Ensuring that any residual materials, particularly hazardous substances, are disposed of properly according to environmental regulations.

International and national regulations play a crucial role in governing the responsible management of hazardous materials in the photovoltaic industry. Many jurisdictions have established guidelines for the environmentally sound disposal and recycling of solar panels, emphasizing the need for proper waste management practices and the reduction of environmental impacts.

As the solar industry continues to grow, ongoing research and innovation in recycling technologies will be essential to improve the efficiency and sustainability of the photovoltaic panel life cycle. Implementing robust recycling practices not only minimizes the environmental footprint of solar energy but also contributes to the development of a circular economy for renewable technologies.

9.1.2. Wind Turbine Safety during Construction and Operation

Ensuring worker safety and the success of wind energy projects depend heavily on wind turbine safety, both during construction and operation. To prevent accidents and injuries during the construction of wind turbines, strict safety protocols are necessary. These procedures are complex and frequently take place in difficult environments. Workers perform jobs like electrical wiring, tower assembly, and turbine installation during construction, frequently at heights of several stories. Strict adherence to safety procedures, the use of personal protective equipment (PPE), and thorough training are necessary to reduce the risks connected with working in high-altitude and dynamic environments (Figure 9.2).

Figure 9.2. *Safety precautions for wind turbine.*

Source: Firoozi AA, Hejazi F. Innovations in Wind Turbine Blade Engineering: Exploring Materials, Sustainability, and Market Dynamics. Sustainability. 2024; 16(19):8564. https://doi.org/10.3390/su16198564.

Once operational, wind turbines continue to pose safety considerations, and regular maintenance activities are crucial to their efficient and safe functioning. Key aspects of wind turbine safety include:

1. **Height Safety and Climbing Procedures:** Wind turbine technicians regularly ascend tall towers for maintenance and repairs. Proper training in working at heights and adherence to strict climbing procedures are imperative to prevent falls and accidents.
2. **Blade Maintenance and Inspection**: Wind turbine blades are subject to wear and tear from environmental factors. Routine inspection and maintenance, including repairs or replacement of damaged blades, are essential to prevent structural failures and ensure safe operation.
3. **Electrical Safety:** Wind turbines generate electricity, and technicians must follow strict electrical safety protocols to avoid electrocution or electrical accidents during maintenance activities.
4. **Fire Safety**: Turbine nacelles contain complex electrical systems and lubricants that pose a fire risk. Adequate fire suppression systems and emergency response plans are critical to managing the risk of fires in wind turbines.
5. **Ice and Cold Weather Precautions:** In cold climates, ice accumulation on blades can impact turbine performance and safety.

De-icing systems and protocols for identifying and mitigating ice-related risks are essential.

6. **Emergency Response Planning:** Comprehensive emergency response plans, including rescue procedures for technicians working at heights, are vital to ensuring a swift and effective response in the event of accidents or failures.
7. **Training and Certification:** Wind turbine technicians must undergo thorough training and certification programs to ensure they are equipped with the knowledge and skills needed to work safely and efficiently.

Regulatory bodies often establish industry standards and guidelines to govern wind turbine safety. Companies involved in wind energy projects are expected to comply with these standards, and ongoing advancements in safety technology and practices contribute to continuous improvement in the sector. As the wind energy industry expands, maintaining a strong safety culture through training, technology innovation, and adherence to best practices is fundamental to the sustainable growth of wind power and the protection of those working in the field.

9.1.3. Wildlife Impact Assessment in Renewable Energy Projects

To ensure responsible and sustainable practices that minimize negative effects on nearby ecosystems, conducting a thorough wildlife impact assessment is a crucial step in the development of renewable energy projects. Understanding and reducing possible effects on wildlife are crucial parts of project planning and environmental stewardship in the context of wind, solar, and hydropower projects. For instance, when using wind energy, turbine blade collisions with birds and bats must be carefully considered. To identify migration patterns and high-risk areas, pre-construction studies frequently use radar technologies and ornithological surveys. This allows developers to put protective measures like curtailment strategies, proper siting, or technological innovations in place to lower the risk of collisions. Similar to this, solar energy projects may have an effect on the biodiversity of the area, especially on species that live on the ground, which calls for habitat assessments and conservation efforts. River ecosystem assessments are necessary for hydropower projects to address possible effects on fish migration, water quality, and aquatic habitats. Developers can implement mitigation measures, habitat restoration, and adaptive management strategies to promote peaceful coexistence between renewable energy development and wildlife conservation by incorporating these assessments into the project lifecycle. In addition to being necessary to comply with regulations, these assessments are essential to fostering long-term environmental sustainability and winning over stakeholders, such as local communities and conservation organizations (Buonocore et al., 2015; Cacciapaglia et al., 2018).

9.1.4. Geothermal Energy Exploration and Seismic Safety

Because of its connection to tectonic activity, geothermal energy exploration is inherently linked to seismic safety concerns, despite its potential as a clean and sustainable energy source. Geological features that are vulnerable to seismic activity are frequently found in areas with geothermal reservoirs, which extract heat from the Earth's interior. Geothermal fluid extraction can change subsurface pressure conditions, which may cause earthquakes. Strong seismic safety precautions are therefore essential to the design and management of geothermal projects. To identify potential seismic hazards and provide information for project planning, thorough site assessments are carried out, which include geological surveys and seismic risk analyses. To reduce induced seismicity, fluid injection and extraction during geothermal operations are carefully controlled. To quickly identify and react to any seismic activity, early warning systems and real-time seismic monitoring are used. Seismologists, regulatory agencies, and geothermal developers must work together to establish and implement protocols that give equal weight to seismic safety and energy production. Although geothermal energy provides a low-carbon alternative, the long-term sustainability and safety of geothermal energy exploration depend on these projects' ability to withstand seismic risks.

9.2. NUCLEAR ENERGY SAFETY

Utilizing the enormous power produced by nuclear reactions requires careful consideration of nuclear energy safety. Nuclear power plants use controlled fission reactions to generate electricity, but because radioactive materials carry inherent risks, every step of the nuclear energy lifecycle requires rigorous safety precautions. To prevent overheating and preserve the integrity of containment structures, nuclear reactors are designed and operated with redundant safety systems. In the event of an emergency, SCRAM emergency shutdown procedures can quickly stop the nuclear chain reaction. Furthermore, by depending on natural processes for stability and cooling, advances in reactor technology, such as passive safety features, increase the resilience of nuclear power plants.

There are many safety concerns associated with the handling and disposal of radioactive waste. Secure storage solutions are needed for long-lived radioactive isotopes. Current research is concentrated on deep geological repositories and sophisticated waste treatment techniques to reduce environmental impact.

Sturdy regulatory frameworks that set safety standards, carry out inspections and monitor the sector to guarantee compliance are enforced by national and international organizations.

International cooperation, facilitated by agencies such as the International Atomic Energy Agency (IAEA), fosters the exchange of knowledge and best practices, strengthening the bond between nations regarding nuclear safety. Open and honest communication with the public is essential to promoting knowledge and confidence in nuclear energy operations. A knowledgeable and involved community is enhanced by public involvement in decision-making procedures and ongoing nuclear safety education.

The operator is responsible for safety is a basic tenet of nuclear power plant operations worldwide. The national regulator is in charge of making sure the licensee operates the plants safely and that the design is accepted. The idea that a regulator's job is to safeguard the environment and people is a second crucial one. National regulators are also in charge of reactor design certification. There are several sets of mechanical codes and standards pertaining to quality and safety, and there are varied degrees of international collaboration among these.

Human performance principles, concepts, and practices have been adopted by the aviation industry, medical field, commercial nuclear power industry, US Navy, DOE, and its contractors, and other high-risk, technologically complex organizations to consciously reduce human error and bolster controls to reduce accidents and events. An estimated 80% of all occurrences are the result of human error. In certain sectors of the economy, this figure approaches 90%. About 20% of incidents include malfunctioning equipment. If the 80% human error rate is dissected further, it becomes clear that most errors related to events are the result of latent organizational weaknesses, while about 30% are the result of a single worker touching the facility's systems and equipment. It is obvious that concentrating efforts on minimizing human error will lessen the chance of occurrences.

9.2.1. Defense-in-Depth

To achieve optimum safety, nuclear plants in the Western world operate using a "defense-in-depth" approach, with multiple safety systems supplementing the natural features of the reactor core. Key aspects of the approach are:

1. High-quality design and construction.
2. Equipment that prevents operational disturbances or human failures and errors from developing into problems.
3. Comprehensive monitoring and regular testing to detect equipment or operator failures.

4. Redundant and diverse systems to control damage to the fuel and prevent significant radioactive releases.
5. Provision to confine the effects of severe fuel damage (or any other problem) to the plant itself.

These can be summed up as prevention, monitoring, and action (to mitigate the consequences of failures).

The safety provisions include a series of physical barriers between the radioactive reactor core and the environment, and the provision of multiple safety systems, each with backup and designed to accommodate human error. As well as the physical aspects of safety, there are institutional aspects that are no less important – see the following section on International Collaboration.

The fuel in a typical plant takes the form of solid ceramic (UO_2) pellets, and during the fuel's burning process, radioactive fission products are mostly contained inside these pellets. To create fuel rods, the pellets are pressed into sealed zirconium alloy tubes. These are housed inside a sizable pressure vessel made of steel with walls as thick as 30 cm. The main water cooling pipework that goes with it is also quite substantial. All of this is contained within a sturdy containment structure made of reinforced concrete with walls that are at least one meter thick. This means that there are three substantial barriers surrounding the fuel, which is stable at extremely high temperatures.

These barriers are monitored continually. The fuel cladding is monitored by measuring the amount of radioactivity in the cooling water. The high-pressure cooling system is monitored by the leak rate of water, and the containment structure by periodically measuring the leak rate of air at about five times atmospheric pressure.

The three basic safety functions in a nuclear reactor are:

- To control reactivity.
- To cool the fuel.
- To contain radioactive substances.

The main safety features of most reactors are inherent – negative temperature coefficient and negative void coefficient. The first means that beyond an optimal level, as the temperature increases the efficiency of the reaction decreases (this in fact is used to control power levels in some new designs). The second means that if any steam has formed in the cooling water there is a decrease in moderating effect so that fewer neutrons are able to cause fission and the reaction slows down automatically.

In the 1950s and 1960s some experimental reactors in Idaho were deliberately tested to destruction to verify that large reactivity excursions were self-limiting and would automatically shut down the fission reaction. These tests verified that this was the case.

Beyond the control rods which are inserted to absorb neutrons and regulate the fission process, the main engineered safety provisions are the back-up emergency core cooling system (ECCS) to remove excess heat (though it is more to prevent damage to the plant than for public safety) and the containment.

Traditional reactor safety systems are "active" in the sense that they involve electrical or mechanical operation on command. Some engineered systems operate passively, for example pressure relief valves. Both require parallel redundant systems. Inherent or full passive safety design depends only on physical phenomena such as convection, gravity or resistance to high temperatures, not on the functioning of engineered components. All reactors have some elements of inherent safety as mentioned above, but in some recent designs the passive or inherent features substitute for active systems in cooling, so forth. Such a design would have averted the Fukushima accident, where the loss of electrical power resulted in the loss of cooling function (Buonocore et al., 2015; Cacciapaglia et al., 2018).

9.2.2. Nuclear Reactor Safety and Emergency Preparedness

Key safety features include:

1. **Reactor Core Design**: Modern reactor designs incorporate safety features such as passive cooling systems, which rely on natural processes to remove heat in the event of an emergency. These designs aim to enhance the inherent safety of reactors.
2. **Control Systems**: Precise control systems continuously monitor and regulate reactor power levels to prevent overheating and maintain stable operating conditions. Emergency shutdown systems, known as SCRAM, provide a rapid and automatic means to shut down the reactor in the event of abnormalities.
3. **Containment Structures**: Robust containment structures are in place to confine radioactive materials in case of a severe accident. This prevents the release of radioactive substances into the environment.
4. **Emergency Cooling Systems**: Redundant emergency cooling systems are essential to prevent the overheating of the reactor core. These systems ensure a continuous removal of heat, even in the absence of regular cooling mechanisms.

Emergency preparedness involves comprehensive planning and training to address potential incidents and protect public safety. This includes:

1. **Emergency Response Plans**: Nuclear power plants develop and regularly update emergency response plans in collaboration with local authorities. These plans outline procedures for notifying the public, evacuating affected areas, and coordinating emergency response efforts.
2. **Drills and Training**: Regular drills and training exercises involving plant personnel, local emergency services, and communities help ensure a coordinated and effective response in the event of an emergency. These exercises simulate various scenarios to test the readiness of both plant operators and external responders.
3. **Public Communication**: Open and transparent communication with the public is crucial during emergency situations. Nuclear power plants provide timely and accurate information to keep the public informed and alleviate concerns.
4. **Off-Site Emergency Centers**: Off-site emergency centers are established to coordinate response efforts beyond the plant boundaries. These centers facilitate communication and collaboration among various stakeholders involved in emergency management (Table 9.1).

Table 9.1. *Comparative Analysis of Energy Policies and Their Impact On Environmental Sustainability Across Regions and Countries*

Region/Country	Energy Policy/Strategy	Key Findings/Output
ASEAN region	Generation expansion planning	Emphasis on expanding energy generation capacities.
Asia–Pacific top consumers	Nuclear energy consumption	The sustainability alignment of nuclear energy is ambiguous.
Baltic states	Biomass and biofuels emphasis	Introduced a scoring system prioritizing biomass and biofuels.
Developed vs. developing countries	Varied energy policies	Socio-economic contexts influence policy design.
European Union (EU)	Trade policy and RE consumption	Policies influence environmental footprint (EFP).

Region/Country	**Energy Policy/Strategy**	**Key Findings/Output**
	RE Policy	The policy has led to significant achievements in RE adoption. In 2020, the share of gross final energy consumption from RESs reached 22%, exceeding the 20% target set for 2020.
	RE and EE	Leading the transition towards renewables and efficiency.
GCC Region	Environmental and energy policies for sustainable consumption and production	Institutional actions are not sufficient for sustainable consumption and production; there is a need for economic diversification and a shift toward RE.
	RE, EE improvements, technological advancements, policy reforms	There has been an increased use of RES, reduction in energy consumption and carbon emissions, and advancements in EE and technology, but overall RE contribution remains low compared to conventional sources.
IEA member countries	Green energy innovations and R&D	Energy R&D is pivotal in reducing emissions.
MENA Region	RE technologies	Positive effect on sustainability development: high cultural acceptance and readiness to invest in clean energy.
	Factors driving RE deployment (economic, financial, political)	Innovation and governance quality are the main drivers; there are calls for sustainable policy options to increase innovation performance.
	Environmental Energy Security Index (EESI)	There is variation in energy security levels and recommendations for investing in local resources, diversifying the energy mix, and energy-efficient technologies.

Region/Country	**Energy Policy/Strategy**	**Key Findings/Output**
	RE, energy intensity, green innovation	RE significantly reduces carbon emissions and emphasizes the need for green innovation and energy-efficient technologies.
	RE investments, government effectiveness	Government effectiveness and RE contribute to carbon neutrality; economic growth initially delays it.
OECD countries	Flexible energy regulations	Policies' effectiveness, cost-efficiency, and design features.
South Asian countries	Energy Security and Environmental Sustainability Index (ESESI)	Scoring system categorizing nations on energy security and environmental sustainability.
Various Nations	Energy Planning Models	Effective policy decision making has no direct tie to environmental sustainability.
Austria	Energy transition strategies towards zero-carbon energy sources	Demonstrated reduced GHG emissions.
Bangladesh	RE sourcing	Strategies tailored to national resources and contexts.
	Solar home systems	Provided electricity to over 5 million households, reducing kerosene usage and improving indoor air quality.
Brazil	Ethanol fuel policies	The policy has led to ethanol meeting 50% of light vehicle fuel needs, contributing to reduced urban air pollution and emissions.
United Kingdom	Carbon pricing vs. renewable subsidies	Carbon pricing was found to be more effective.
	RE sourcing	Strategies tailored to national resources and contexts.
China	RE-focused policies	Emphasis on creating a low-carbon society.
	Top-1000 Enterprises Energy-Saving Program	Analysis of policies targeting SO_2 emissions.

Region/Country	Energy Policy/Strategy	Key Findings/Output
	Hydropower integration	Utilized hydropower potential.
Croatia	Energy transition strategies	Demonstrated reduced GHG emissions.
Germany	Sustainable energy transition strategies	Policies focus on social equity and implementation challenges.
	Energy transition strategies to increase energy access and affordability, as well as diversification towards renewable energy	Demonstrated reduced GHG emissions.
	RE policies (feed-in tariffs, RPS)	Policies effective in promoting RE.
	Carbon pricing vs. renewable subsidies	Carbon pricing is more effective.
Greece	RE within the EKC framework	Aligned with EU's 2030 targets and European Green Deal Investment Plan.
India	National solar mission	India's solar power capacity grew from 2.6 GW in 2014 to over 40 GW by 2019, reducing coal dependence and emissions.
	RE-focused policies	Emphasis on creating a low-carbon society.
Italy	Sustainable energy transition strategies	Policies focus on social equity and implementation challenges.
	RE within the EKC framework	Aligned with EU's 2030 targets and European Green Deal Investment Plan.
Japan	Sustainable energy transition strategies	Policies focus on social equity and implementation challenges.
	RE-focused policies	Emphasis on a low-carbon society.
Kenya	RE sourcing	Strategies tailored to national resources and contexts.
Norway	Hydropower integration	Utilized hydropower potential.
Portugal	RE within the EKC framework	Aligned with EU's 2030 targets and European Green Deal Investment Plan.

Region/Country	Energy Policy/Strategy	Key Findings/Output
South Korea	RE-focused policies	Emphasis on creating a low-carbon society.
Spain	RE within the EKC framework	Aligned with EU's 2030 targets and European Green Deal Investment Plan.
Sweden	Hydropower integration	Utilized hydropower potential.
United States	RE policies (feed-in tariffs, RPS)	Policies effective in promoting RE.
	Renewable portfolio standards	The policy has boosted RE capacity and generation, leading to reduced emissions and health benefits.

Source: Elkhatat A, Al-Muhtaseb S. Climate Change and Energy Security: A Comparative Analysis of the Role of Energy Policies in Advancing Environmental Sustainability. Energies. 2024; 17(13):3179. https://doi.org/10.3390/en17133179.

9.2.3. Radioactive Waste Management and Disposal

The production of electricity generates waste, just like any other industry. Regardless of the fuel used, waste generated during the production of electricity needs to be managed to protect public health and reduce environmental impact. This entails separating or diluting radioactive waste in a way that ensures any radionuclides that are released back into the biosphere will do so at a rate or concentration that is safe. To do this, almost all radioactive waste is managed and contained, though some obviously need to be buried deep and permanently. In contrast to all other forms of thermal electricity generation, all waste from nuclear power generation is regulated and cannot lead to pollution. The very large amount of energy generated from a very small amount of fuel and the comparatively small amount of waste produced during this process are the defining characteristics of nuclear power. Nonetheless, a large portion of the waste generated is radioactive, necessitating careful handling as a hazardous material. Radioactive waste is produced during every stage of the nuclear fuel cycle, and the expense of handling and getting rid of it is included in the price of power that is, it is internalized and funded by the people who use power. Not just radioactive waste, but all toxic waste must be managed safely; in nations that use nuclear energy, radioactive waste makes up a very small percentage of the total amount of industrial hazardous waste produced. The nuclear fuel cycle is not the only source of radioactive waste. In the fields of medicine, agriculture, research, manufacturing, non-destructive testing, and mineral exploration, radioactive materials are widely utilized. All radioactive waste has a decreasing level of hazard over time due to its radioactivity, unlike other hazardous industrial materials.

9.2.3.1. Types of Radioactive Waste

Any material deemed to have no further use that is radioactively contaminated or intrinsically radioactive is considered radioactive waste. Certain materials, like spent nuclear fuel and plutonium, are classified as waste according to government policy. A radionuclide's half-life is the amount of time it takes for half of its atoms to decay and, consequently, for half of its radioactivity to dissipate. Long half-lived radionuclides are more likely to emit alpha and beta radiation, which facilitates handling, whereas short half-lived radionuclides are more likely to emit gamma rays, which are more invasive. All radioactive waste eventually decomposes into elements that are not radioactive. A decaying isotope decays more quickly the more radioactive it is. Depending mainly on its radioactivity level, radioactive waste is usually categorized as low-level (LLW), intermediate-level (ILW), or high-level (HLW).

9.2.3.1.1. Low-Level Waste

The radioactive content of low-level waste (LLW) is limited to four giga-becquerels per ton (GBq/t) of alpha activity or 12 GBq/t of beta-gamma activity. LLW is appropriate for disposal in facilities close to the surface and doesn't require shielding during handling or transportation. Hospitals, businesses, and the nuclear fuel cycle all produce LLW. Paper, rags, clothes, tools, filters, and other items are included, which are mostly composed of short-lived radioactive elements in trace amounts. Before being disposed of, LLW is frequently compressed or burned to minimize its volume. Of all radioactive waste, LLW makes up 90% of the volume but only 1% of its radioactivity.

9.2.3.1.2. Intermediate-Level Waste

Although intermediate-level waste (ILW) produces less heat than LLW (<2 kW/m^3), it is still more radioactive than LLW and should not be considered when designing or choosing storage and disposal facilities. ILW needs to be shielded since its radioactivity levels are higher. ILW generally consists of contaminated materials from reactor decommissioning, resins, chemical sludges, and metal fuel cladding. For disposal, smaller items and any non-solids can be solidified in bitumen or concrete. Roughly 7% of the total volume and 4% of the radioactivity of all radioactive waste is found in it.

9.2.3.1.3. High-Level Waste

High-level waste (HLW) has decay heat (>2kW/m^3) that is sufficiently radioactive to raise both its own and its surroundings' temperatures noticeably. HLW needs cooling and shielding as a result. The "burning" of uranium fuel in a nuclear reactor produces heavy lead oxide (HLO). The transuranic elements and fission products produced in the reactor core are contained in HLW. Only 3% of the volume and 95% of the radioactivity in produced waste is accounted for by HLW.

There are two distinct kinds of HLW:

- Used fuel that has been designated as waste.
- Separated waste from reprocessing of used fuel.

Depending on how long it will take for a given radionuclide's radioactivity to drop to levels deemed safe for humans and the environment, HLW can contain both long-lived and short-lived components. In general, the management and disposal of heavy HLW depends on the ability to distinguish between short-lived fission products and long-lived actinides. Since HLW is the subject of much discussion in relation to nuclear power, it is handled appropriately.

9.2.3.1.4. Very Low-Level Waste

Radioactive materials are present in very low-level waste (VLLW) and exemption waste at levels deemed safe for humans and the environment. The majority of its components are demolished materials, including bricks, concrete, plaster, metal, valves, and plumbing generated during nuclear industrial sites' restoration or dismantling processes. Other industries: steel, chemicals, food processing, so forth additionally generate VLLW due to the natural radioactivity concentration found in specific minerals used in their production processes. Hence, the waste is disposed of with household garbage, though nations like France are currently building VLLW disposal facilities that are especially intended for this purpose.

9.2.3.2. Storage and Disposal

Waste can be stored at any point during the management procedure. To store waste properly, it must be kept separate from the outside world and maintained in a retrievable manner. To facilitate the management of the following phase, waste can be stored (e.g., by allowing its natural radioactivity to decay). Storage facilities can be found separately from the facility where it was produced, but they are typically located on site at the power plant. Waste is disposed of when its use is no longer anticipated, and in the case of heavy lithium waste (HLW), this occurs when radioactivity has decreased to comparatively low levels after 40–50 years.

9.2.3.2.1. LLW and Short-Lived ILW

Most LLW and short-lived ILW are typically sent to land-based disposal immediately following packaging. This means that for the majority (>90% by volume) of all of the waste types, a satisfactory disposal means has been developed and is being implemented around the world.

Some low-level liquid waste from reprocessing plants is discharged to the sea. This includes radionuclides which are distinctive, notably technetium-99 (sometimes used as a tracer in environmental studies), and this can be discerned many hundred kilometers away. However, such discharges are regulated and controlled, and the maximum radiation dose anyone receives from them is a small fraction of natural background radiation.

Small amounts of radioactive gases are released by nuclear power plants and reprocessing facilities (e.g., trace levels of iodine-131 as well as krypton-85 and xenon-133 to the atmosphere. All three gases have brief half-lives, xenon-133 and krypton-85 are chemically inert, and postponing the release of the emissions reduces their radioactivity. The overall impact is too negligible to be taken into account in a life-cycle analysis. Additionally, a small amount of tritium is produced, although regulators do not view this release as noteworthy.

9.2.3.2.2. Long-Lived ILW and HLW

The idea of deep geological disposal has gained widespread acceptance due to the lengthy radioactive half-lives of some HLW and ILW, including spent fuel when regarded as waste. Although many different long-term waste management strategies have been studied, deep disposal in a mined repository is currently the method of choice in the majority of nations. The United States is currently using the Waste Isolation Pilot Plant (WIPP) deep geological waste repository to dispose of transuranic waste, or long-lived, plutonium-contaminated ILW from military sources.

As of yet, final HLW repositories have not been practically necessary. Used fuel can be disposed of directly or reprocessed, as previously mentioned. In any case, there is a compelling technical argument to postpone HLW final disposal for 40 to 50 years following removal, by which time radioactivity and heat will have decreased by more than 99%. Used fuel is typically temporarily stored in ponds connected to individual reactors, in common pools at multi-reactor sites, or sporadically at a central location. Approximately 263,000 tons of spent fuel are being stored at the moment. With a growing percentage in dry storage, more than two thirds of this is in storage ponds.

Reactor storage ponds and those at centralized facilities are 7–12 meters deep, allowing several meters of water to cover the spent fuel, which is assembled into racks that are usually 4 meters long and stand upright. The several racks are composed of metal and have built-in neutron absorbers. The water in circulation cools and protects the fuel. These are sturdy pools with steel liners made of thick reinforced concrete. Reactor ponds are frequently built to hold all of the spent fuel generated during the reactor's anticipated operating lifetime.

Certain fuels are kept in dry casks or vaults with air circulation inside concrete shielding after cooling in ponds for a minimum of five years. Multi-purpose canisters (MPCs), or sealed steel casks, are a common system that can hold up to 40 fuel assemblies with inert gas each. Additionally, used fuel can be transported and eventually disposed of using casks or MPCs. Each is housed in an air-conditioned concrete and steel storage module for storage. They can be below grade with only the tops visible, or they can be standing on the surface at a height of roughly 6 meters and being cooled by air convection. The modules offer complete shielding and are sturdy. The heat load of each cask can reach 45 kW.

Reprocessing spent reactor fuel necessitates solidifying the liquid HLW. In addition, the HLW produces a significant amount of heat and needs to be cooled. It is sealed into large stainless steel cylinders that are roughly 1.3 meters high, vitrified into borosilicate (Pyrex) glass, and kept underground for possible disposal. This material is considered waste since there is no way that it could be used in the future. In addition to the two operating in France, there are plants in the U.K and Belgium that vitrify leftover heavy fuel waste. Some of these Western European plants have been in operation for 30 years, and their annual capacity is 2500 canisters (1000 t). More than 6000 canisters of vitrified HLW have been produced by Sell Afield, U.K. Used reactor fuel will still contain all of the extremely radioactive isotopes if it is not further processed. Fuel waste that isn't recycled is disposed of directly as HLW.

It also produces a lot of heat and needs to be cooled. Nonetheless, it represents a potentially valuable resource because it mostly consists of uranium (with a small amount of plutonium), and there is growing resistance to disposing of it irretrievably. For final disposal, to ensure that no significant environmental releases occur over tens of thousands of years, "multiple barrier" geological disposal is planned. This technique will immobilize the radioactive elements in HLW and long-lived ILW, and isolate them from the biosphere. The multiple barriers are:

1. Immobilization of waste in an insoluble matrix such as borosilicate glass or synthetic rock (fuel pellets are already a very stable ceramic, UO_2).

2. Contain waste sealed inside a corrosion-resistant container, such as stainless steel.
3. Isolate waste from people and the environment, so eventually locate it deep underground in a stable rock structure.
4. Delay any significant migration of radionuclides from the repository, so surround containers with an impermeable backfill such as bentonite clay if the repository is wet (Buonocore et al., 2015; Cacciapaglia et al., 2018).

9.3. FOSSIL FUELS AND ENVIRONMENTAL IMPACT

Fossil fuel extraction, processing, and use have a substantial negative impact on the environment and contribute to both ecological degradation and climate change. Fossil fuel combustion releases enormous amounts of carbon dioxide, which is the main cause of the greenhouse effect and exacerbates global warming. Furthermore, methane leakage, a potent greenhouse gas with immediate environmental impact, is a risk associated with the extraction of natural gas, a fossil fuel that is frequently promoted as a cleaner alternative. Beyond issues with the atmosphere, the use of fossil fuels is associated with air pollution, which releases nitrogen oxides and particulate matter and has a negative impact on public health and air quality. Accidents involving the transportation and extraction of oil can result in spills that damage terrestrial and marine environments, making cleanup efforts difficult to overcome. Moreover, habitat disruption from fossil fuel infrastructure and changes in land use, particularly in surface coal mining, are factors in the decline of biodiversity and fragmentation of ecosystems. The industry's high water consumption and the potential for water contamination—especially during the fracking process—highlight the intricate web of environmental issues connected to the use of fossil fuels. To mitigate the extensive environmental toll caused by fossil fuel activities, addressing these issues requires a comprehensive shift to cleaner energy sources and the implementation of strict environmental policies.

9.3.1. Environmental Impacts

Environmental impacts are those concerned with the built and natural environment, including air, water, and biodiversity. Events cause impacts resulting from activities such as construction, energy use and transport of spectators and participants, which in turn impact on major environmental issues such as climate change and the reduction in natural resources.

9.3.1.1. Integrating Environmental Considerations

An event has effects on the environment both before and after it happens, as well as during its preparation. The negative environmental impact increases with the size of the event for many types of impacts. The kinds of environmental effects that result from an event can be greatly influenced by decisions taken during the early stages of planning. To deliver the event with environmental sustainability at the center of all decision-making, it is therefore best practice to establish a baseline environmental assessment at the beginning of the preparation phase. Environmental policies must consider supplier behavior as well; this can be done, for instance, by implementing green procurement guidelines.

9.3.1.2. Possible Outcomes

1. **Environmental Quality:** the nature of major events having significant media attention in the build-up to the event, provides an opportunity for them to demonstrate leadership of environmental best practice and standards, such as waste recycling, which may be adopted more widely by other organizations in the host nation after the event.
2. **Climate Action:** Very large scale multinational events with media exposure to a wide audience, also provides an opportunity for positive leadership and action to address climate change, which can in turn support multinational cooperation for this global issue.

9.3.1.3. Inputs

A good practice is for major events to address environmental issues within the sustainability plans at the start of the preparation phase. Key areas to be considered are:

1. **Waste:** One of the most visible environmental impacts arising from a major event. However, volumes, nature, and destination of waste associated with an event, either as part of infrastructure development, event operations or visitor consumption are complex to measure. This is due to the mix of responsible agents from venues, local authorities, organizers, and suppliers, to spectators and participants.

Example KPIs

- Proportion of waste diverted from landfill (including recycling, composted).
- Proportion of food and drink containers made from recyclable/ biodegradable material.

- Proportion of signs, banners, tents, so forth. that can be used for multiple years.

2. **Energy:** Energy use during the preparation and delivery of a major events takes different forms, whether from the construction of venues, energy consumed by spectators travelling to an event, or energy used in running event venues. Critical is to minimize the proportion of this energy developed through the use of fossil fuels due to the impact on climate change from greenhouse gas emissions. The practice of calculating the carbon footprint of an event is becoming more common. This is a useful tool to identify opportunities for avoiding and reducing greenhouse gas emissions. Carbon offsetting can be used to offset any unavoidable emissions through "carbon compensation" schemes.

Example KPIs

- Proportion of energy sourced from renewables or purchased through energy providers' green power programme.
- Quantified emission reductions against baseline carbon footprint.
- Total event carbon footprint (CO_2 equivalent).

3. **Water:** Increasingly water is becoming a scarce resource in many host communities, if only at certain times of the year. Therefore, reducing water consumption is a key issue for sustainable development and an area of focus for major events, both in preparation for, and during, the event.

Example KPIs

- Proportion of water used from sustainable sources, for example rainwater harvesting.

4. **Transport:** Transport can be one of the most complex aspects of hosting a major event and a large contributor to costs, greenhouse gas emissions, and reduced air quality. Encouraging people to use greener transport modes to get to and from events reduces travel time, traffic congestion, greenhouse gas emissions, and parking issues.

Example KPIs

- Proportion of visitor journeys by public transport.
- Proportion of visitor journeys by sustainable travel (e.g., cycling, zero-emission vehicles).

5. **Natural Capital:** Open spaces created as part of new construction

for major events can have a significant impact on local ecosystems. Healthy ecosystems provide numerous benefits, such as amenities, mitigation against urban heat island effects, flood alleviation, soil stability, improved air quality and soundscape, and food production.

Example KPIs

- Environmental safeguards put in place, for example pollution prevention measures.
- Net gain in accessible green space.

9.3.2. Oil Spill Response and Prevention

To lessen the severe effects on the environment and the economy that arise from unintentional oil spills into marine and terrestrial ecosystems, effective response and prevention measures are essential. To reduce the chance of spills, prevention initiatives emphasize strict laws, cutting-edge technologies, and industry best practices. The risk of oil spills during exploration and transportation can be decreased by using cutting-edge drilling and extraction technologies, modernizing vessel design, and providing personnel with extensive safety training. Furthermore, improvements in spill detection systems—like satellite imagery and real-time monitoring—improve early warning systems and enable quick reactions to possible incidents. Strong response plans are necessary because oil spills do happen, even with precautions taken. Booms and other quick containment techniques are essential for stopping the spread of oil and safeguarding sensitive areas. Moreover, the removal or reduction of the effects of oil spills is facilitated by the deployment of specialized vessels and equipment for skimming, *in-situ* burning, and dispersant application. An efficient response requires coordinated efforts from government agencies, business partners, and environmental organizations. Innovative cleanup technology research is still advancing, opening up new possibilities for effective and long-lasting approaches to responding to oil spills. Examples of these technologies include bioremediation and the use of autonomous cars for monitoring and intervention. Campaigns for public awareness are also essential because they inform the public about the significance of promptly reporting oil spills and reducing their effects through responsible behavior. To protect our oceans, coastlines, and ecosystems from the severe and long-lasting effects of oil pollution, we must continue to refine our methods for preventing and responding to oil spills as we navigate a future that calls for increased energy production.

9.3.3. Air Quality and Emissions Control

Environmental stewardship, public health, and the worldwide effort to mitigate climate change all depend critically on maintaining air quality and putting into place efficient emissions control measures. The primary sources of air pollutants, such as particulate matter, nitrogen oxides (NOx), sulfur dioxide (SO_2), and volatile organic compounds (VOCs), are the combustion of fossil fuels, industrial processes, and transportation. These pollutants have far-reaching effects because they cause acid rain, smog, respiratory illnesses, and climate change. Strict emission control laws and air quality standards are put into place globally to address these issues. Modern technologies are essential for cutting emissions.

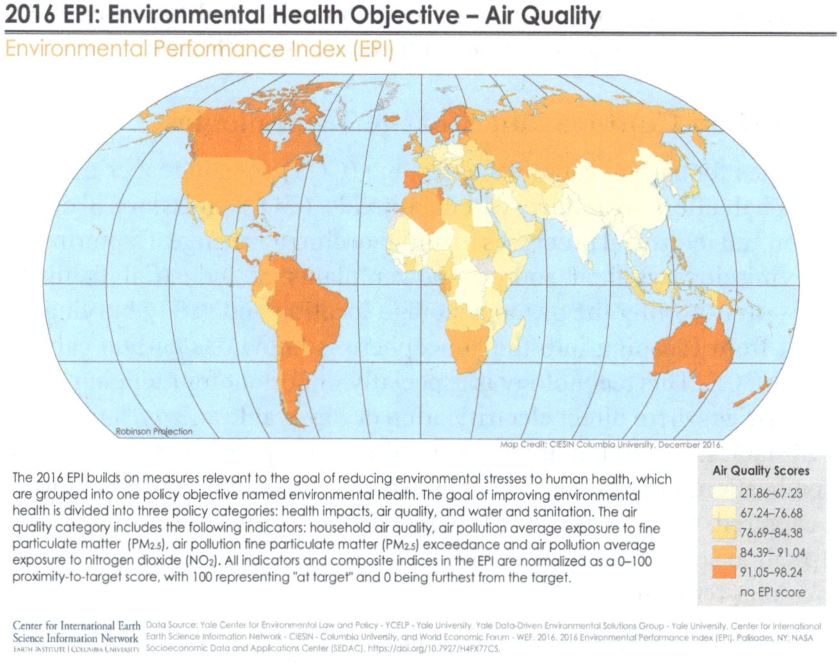

Figure 9.3. *The control of emissions and air pollutants is a key component in reducing environmental impact while staying compliant with government regulations.*

Source: By SEDACMaps – 2016 EPI: Environmental Health Objective – Air Quality, CC BY 2.0, https://commons.wikimedia.org/w/index.php?curid=85086693.

Examples include scrubbers in power plants, electrostatic precipitators in industrial facilities, and catalytic converters in automobiles. Promoting greener energy sources, such as renewable energy and electrification of transportation, helps reduce air pollution even more. Furthermore, reducing emissions is aided by the creation and implementation of sustainable practices like urban planning and green building designs. Modeling and ongoing air quality monitoring assist in determining the efficacy of control measures and provide information for prompt interventions. Since air pollutants frequently cross international borders, international cooperation is crucial to addressing transboundary pollution. The development of policy frameworks that incentivize emissions reductions and the integration of cutting-edge technologies like carbon capture and storage are essential if we are to achieve a sustainable future. To promote a shared commitment to improving air quality, encourage responsible consumption, and support policies that prioritize clean air for both the present and the future generations, public awareness and engagement are also essential. In conclusion, extensive efforts in emissions reduction and air quality management are essential for preserving ecosystems, preserving human health, and lessening the wider effects of air pollution on the climate and environment (Figure 9.3).

9.3.4. Carbon Capture and Storage Technologies

Technologies for carbon capture and storage (CCS) constitute a critical frontier in the global effort to reduce carbon dioxide (CO_2) emissions from power generation and industrial processes to mitigate climate change. Capturing carbon dioxide emissions at their source—power plants or industrial facilities, for example—transporting the gas to a storage location and safely burying it there to keep it from escaping into the atmosphere is known as carbon capture and storage, or CCS. This technology is especially important for reducing emissions from industries where direct electrification or renewable energy sources present major obstacles. Several methods are used in the capture process, such as oxy-fuel combustion, pre-combustion capture, and post-combustion capture. After being captured, the CO_2 is moved to appropriate geological formations—such as deep saline aquifers or depleted oil and gas reservoirs—by pipelines or ships for safe, long-term storage.

In particular, CCS provides an invaluable link between the ongoing use of fossil fuels and the shift to a low-carbon future in industries where it is difficult to completely eliminate emissions. It could help achieve climate targets by cutting greenhouse gas emissions dramatically. However, obstacles to public acceptance, regulations, and the economy stand in the way of widespread adoption. One of the ongoing challenges in implementing CCS is the high infrastructure and technology costs. Furthermore, strong, and transparent regulatory frameworks are necessary to guarantee the deployment of CCS projects in a safe and environmentally responsible manner.

The acceptance and perception of CCS technologies by the general public are also important determinants of their success. Building trust requires open and honest communication about the effectiveness and safety of storage locations as well as involvement with the neighborhood. To overcome these obstacles, governments, businesses, and academic institutions must work together to promote innovation, reduce costs, and provide the required incentives for the deployment of CCS. Hence, CCS can be a key component of a comprehensive plan to combat climate change, helping to achieve significant reductions in emissions in a range of sectors and serving as an essential instrument for decarbonizing the world economy and fulfilling aggressive climate targets (Buonocore et al., 2015; Cacciapaglia et al., 2018).

9.3.5. Occupational Health and Safety in Extractive Industries

Because these jobs are inherently dangerous, occupational health and safety (OHS) is crucial in extractive industries, which include mining, oil, and gas extraction, and quarrying. There are many risks that workers in extractive industries must deal with, such as physical hazards, exposure to toxic substances, and the possibility of accidents that result in serious injuries or fatalities. Strong OHS regulations must be put in place to protect employee health, reduce the risk of occupational illnesses, and stop accidents. This entails wearing personal protective equipment (PPE), adhering to strict safety procedures when operating machinery, and holding frequent training sessions to inform employees about potential risks and safe work procedures.

In addition, there may be long-term health risks associated with dust, noise, and vibration exposure during extractive operations. In the extractive industries, OHS requires careful monitoring and management of these exposures. To reduce the negative effects on employees' health, engineering controls such as noise barriers, ventilation systems, and other measures are put in place. Frequent medical exams and health screenings are crucial for the early identification of occupational illnesses, which enables prompt intervention and treatment. In the event of an accident, prompt, and efficient response is facilitated by comprehensive emergency response plans and first aid training.

Since operations in the extractive industries frequently have an impact on local communities, community engagement is also essential. Fair employment practices, community concerns, and minimizing environmental disruptions are all components of responsible extraction practices. To balance economic development with social and environmental sustainability, industry stakeholders, governments, and local communities work together.

International standards and governmental laws are major influences on how OHS procedures are carried out in the extractive industries. Enforcing adherence to safety standards and holding companies responsible are contingent upon the enforcement of these regulations. The constant exploration and creation of novel technologies, like automation and remote monitoring, helps to lower the amount of risk that employees face in the workplace. To safeguard employees, reduce negative environmental effects, and promote sustainable development, extractive industries must place a high priority on occupational health and safety.

To ensure a safer and more sustainable future for those engaged in extractive activities, industry players, regulatory agencies, and local communities must work together in a proactive and cooperative manner (Buonocore et al., 2015; Cacciapaglia et al., 2018).

9.4. EMERGENCY RESPONSE AND CRISIS MANAGEMENT

To ensure public safety and lessen the effects of unanticipated events, emergency response and crisis management are essential, especially in industries like public infrastructure, energy, and healthcare. Proactive steps to anticipate, address, and recover from a variety of crises, such as natural disasters, industrial accidents, or security breaches, are part of a strong emergency response plan. Creating emergency response teams, coordinating with pertinent authorities, and developing clear communication protocols are all examples of effective emergency response strategies. To evaluate staff readiness and improve response protocols, regular drills and simulations are carried out. Beyond the initial phase of reaction, crisis management focuses on ways to control and recover from a crisis' long-term effects.

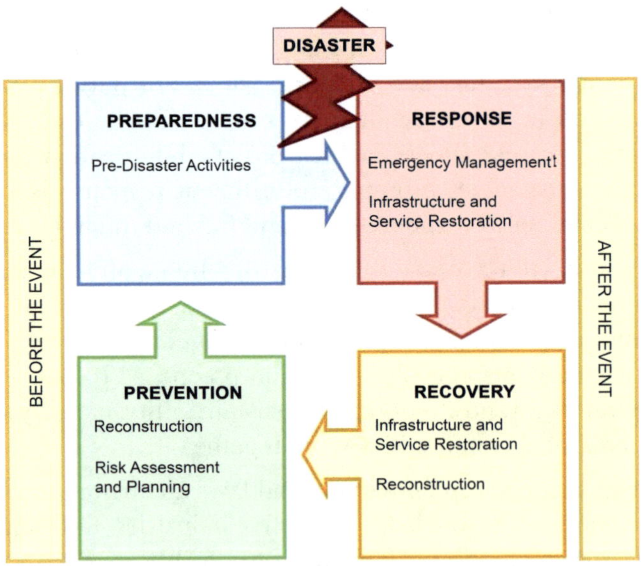

Figure 9.4. *Linking emergency response to crisis management.*

Source: Schwarz K, Aranda DA, Hartmann M. Towards Automated Situational Awareness Reporting for Disaster Management—A Case Study. Sustainability. 2023; 15(10):7968. https://doi.org/10.3390/su15107968.

To guarantee the continuation of vital operations, this entails contacting stakeholders, putting out public relations campaigns, and, when necessary, putting business continuity plans into action. A comprehensive and coordinated response to emergencies requires cooperation between public and private sectors as well as local communities. This highlights the value of shared resources, information sharing, and a cohesive crisis management strategy. The ever-changing global environment necessitates that emergency response and crisis management strategies be flexible and effective to preserve human life, defend infrastructure, and foster resilience in the aggregate (Figure 9.4).

9.4.1. Energy Infrastructure Resilience to Natural Disasters

Ensuring the resilience of energy infrastructure to natural disasters is a critical imperative to maintain the reliability and functionality of power systems in the face of extreme environmental events. Hurricanes, earthquakes, floods, and wildfires are among the natural disasters that can pose substantial threats to energy infrastructure, potentially leading to widespread outages and long-term disruptions. Resilience strategies involve a multifaceted approach, encompassing both proactive measures and adaptive responses:

1. **Infrastructure Design and Siting**: Strategic planning involves considering the susceptibility of potential sites to natural disasters when designing and locating energy infrastructure. Implementing elevated or fortified structures in flood-prone areas, earthquake-resistant designs, and wildfire-resistant materials can enhance the resilience of the infrastructure.
2. **Redundancy and Backup Systems**: Integrating redundancy into energy systems, including backup power sources and alternative routes for energy transmission, helps mitigate the impact of infrastructure failures during disasters. Well-designed backup systems can ensure continued energy supply even when primary systems are compromised.
3. **Advanced Monitoring and Early Warning Systems**: Leveraging advanced monitoring technologies, such as sensors and satellite imagery, enables real-time assessment of changing environmental conditions. Early warning systems provide crucial time for operators to implement shutdown procedures, initiate backup systems, or take other preventive measures in anticipation of a natural disaster.
4. **Community Engagement and Education**: Educating local communities about disaster preparedness and resilience measures

fosters a collaborative approach to managing the impact of natural disasters on energy infrastructure. Community engagement can include the development of emergency response plans, public awareness campaigns, and drills to enhance readiness.

5. **Adaptive Planning and Response**: Dynamic response plans that can be adapted to different disaster scenarios are essential. This involves ongoing training for emergency response teams, regular drills, and the ability to mobilize resources quickly to address unforeseen challenges during and after a natural disaster.
6. **Regulatory Standards and Compliance**: Governments and regulatory bodies play a crucial role in establishing and enforcing standards that enhance the resilience of energy infrastructure. Compliance with these standards ensures that energy facilities are designed and maintained to withstand the impact of natural disasters.

As the frequency and intensity of natural disasters continue to rise, prioritizing the resilience of energy infrastructure is essential for maintaining the stability of power systems, safeguarding communities, and supporting overall societal well-being.

9.4.2. Crisis Communication Strategies in Energy Incidents

When it comes to energy incidents, crisis communication strategies are essential for handling the fallout and influencing public opinion. Unexpected events like power outages, industrial accidents, or environmental emergencies can have significant repercussions. Transparent and quick information sharing is essential for addressing concerns and stopping the spread of false information. It gives stakeholders and the general public timely updates. A coordinated and consistent response is ensured by establishing clear communication protocols and assigned spokespeople, which avoids misunderstanding and upholds credibility. Utilizing a variety of communication channels, such as social media, official websites, and traditional media, enables wide outreach while taking audience preferences into account. A sense of accountability and empathy are fostered by acknowledging and responding to the public's worries and inquiries, whether through hotlines, town hall meetings, or social media engagement. Consistent updates and prompt reactions to new developments show a dedication to managing expectations and informing stakeholders. Training in readiness is essential because it gives communication teams the tools they need to handle stressful situations well. Beyond quick fixes, community involvement becomes essential because it fosters cooperative crisis management by comprehending and taking into account the viewpoints of nearby communities. By using post-incident evaluations to learn from previous incidents, organizations can continuously improve their communication strategies and prepare for future crises. Last but not least, flexible messaging that acknowledges the individuality of every incident allows organizations to customize their communication to meet particular needs and circumstances, enhancing the applicability and efficacy of their crisis communication initiatives. To put it simply, an all-encompassing crisis communication plan for energy-related incidents incorporates accuracy, empathy, adaptability, and transparency to build trust and resilience in the face of unforeseen difficulties.

9.5. RENEWABLE ENERGY TRANSITION CHALLENGES: HARNESSING INNOVATION FOR A SUSTAINABLE FUTURE

Although it offers a low-carbon, sustainable future, the world's shift to renewable energy sources is beset with a number of technical, financial, and socio-political difficulties. Renewable energy sources, like wind and solar, are erratic, which makes grid stability extremely difficult to achieve. To store excess energy for use during low-generation times, energy storage technologies must advance. Large-scale energy storage systems, like advanced batteries and grid-scale storage solutions, require significant investment and innovation to achieve this. In addition, the decentralized and variable nature of renewable energy necessitates a significant modernization of the current energy infrastructure, which was largely built for the centralized generation of fossil fuels. The economic environment comes with its own set of difficulties, such as the requirement for large upfront investments in renewable projects and the difficulties of moving away from well-established, frequently supported fossil fuel industries. The development of renewable energy can be hampered by inconsistent or insufficient regulations, and policy frameworks are important. To draw investments and develop a market that is favorable for renewable energy, a stable and encouraging policy environment is essential. This environment should include incentives, subsidies, and unambiguous regulatory guidelines. For renewable energy projects to be successfully deployed, community involvement and overcoming opposition are crucial, and social acceptance is crucial. International cooperation, technology transfer, and financial assistance are needed to close the gap on global energy inequality, where wealthy nations achieve notable progress while developing nations struggle with energy poverty. The availability and affordability of critical technologies are impacted by issues with sustainability, ethical sourcing, and geopolitical tensions in the renewable energy supply chain. The workforce transition is a complex issue that calls for the fair transition of workers from traditional energy sectors as well as the acquisition of new skills. Governments, business leaders, and the general public must work together to address these issues to enable a seamless and inclusive transition to a future of renewable energy that is socially just and environmentally sustainable (Figure 9.5).

Figure 9.5. *Renewable energy transition challenges.*

Source: By USPS, scanned by Greudin – USPS, Public Domain, https://commons.wikimedia.org/w/index.php?curid=70894.

9.5.1. Challenges in the Renewable Energy Transition

As the world races towards a sustainable future, the transition to renewable energy sources has gained unprecedented momentum. Yet this journey is not without challenges. Some of the challenges faced in the transition to renewable energy are mentioned below, along with their solutions.

9.5.1.1. Intermittency

One of the most prominent challenges to the adoption of renewable energy is the intermittent nature of renewable sources like solar and wind. The sun doesn't always shine, and the wind doesn't always blow. This variability can disrupt the stability of the power grid. However, energy storage solutions are rapidly evolving to address this issue. By foreseeing when renewable sources will be at their peak, downtime can be minimized, and backup power sources can be activated seamlessly.

9.5.1.2. Storage Limitations

Energy Storage may be an important innovation, but it also has limitations. When the sun is abundant or the winds are strong, the excess energy generated needs to be stored. Lithium-ion and other advanced battery technologies offer a promising solution to intermittency challenges. These batteries can store excess energy during peak production times and release it when demand is high, helping to balance the grid.

9.5.1.3. Infrastructure Upgrades

Transitioning to renewable energy requires more than just installing solar panels or wind turbines. Old infrastructure poses a significant challenge, as it may not be optimized for efficient renewable energy distribution. Upgrade grid infrastructure to accommodate higher renewable energy penetration, incorporating advanced grid management and energy storage technologies to ensure stability and reliability.

9.5.1.4. Innovations and Technological Advancements in Renewable Energy

Renewable energy sources and their technological advancements play a key role in shifting our global energy landscape towards sustainability and reducing our reliance on fossil fuels, thereby mitigating climate change.

9.5.1.5. Solar Energy Dominance

The renewable energy revolution is still led by solar energy, and photovoltaic technology is getting more efficient and affordable. There has been a notable increase in solar installations worldwide in recent times. Approximately 6000–7000 MW of home solar power will be added annually between 2023 and 2027. Innovations that maximize energy capture and blend in seamlessly with urban architecture, like transparent solar panels integrated into building materials, are also gaining popularity.

9.5.2. Microgrids: Empowering Local Energy Solutions

A microgrid is a small-scale energy system that generates, distributes, and stores electricity. It can function either independently or in tandem with the main grid. By supplying dependable electricity, encouraging self-sufficiency, and supporting communities in difficult grid environments, it changes lives and economies. Microgrids boost the economy and provide electricity to isolated places. Additionally, they increase energy resilience by maintaining critical infrastructure, such as hospitals and communication centers, by cutting off from the main grid in the event of a disaster or other failure.

9.5.2.1. Microgrids Market Size

By 2032, the size of the global microgrid market is predicted to have surpassed USD 168.64 billion, growing at a compound annual growth rate (CAGR) of 18% between 2023 and 2032.

The onshore and offshore wind energy markets are expanding at a rapid pace. Technological developments in wind turbines have led to lower costs and higher energy output, thanks to things like bigger, more effective blades. In deeper waters where conventional offshore wind farms are impractical, floating wind farms have emerged as a potentially viable solution for wind resource harvesting. According to wind power specialists at Renewable Energy World, costs for wind energy could drop by as much as 35% by 2035.

9.5.2.2. Mart Grids: Empowering Energy Efficiency

Smart grids represent a radical change in the way we produce, distribute, and use energy—they are more than just a technical advancement. Encouraging energy efficiency is one of smart grids' main purposes. By enabling demand-response mechanisms, they give utilities the ability to encourage customers to use less energy during peak hours, stabilizing the grid and averting overloads.

9.5.3. Fossil Fuel Facilities and Safety Considerations

Facilities that use fossil fuels, such as oil refineries, natural gas processing facilities, and coal-fired power plants, present special safety challenges because of the materials they use and the intricate procedures they use. Because of the handling and processing of volatile, toxic, and flammable materials that these facilities engage in, strict safety regulations are required to safeguard both the workers and the surrounding communities.

9.5.3.1. Significant Risk of Fires and Explosions

Strict fire prevention measures, including fire-resistant materials, advanced fire detection systems, and emergency response plans, are crucial to mitigate these hazards.

9.5.3.2. Chemical Exposure and Toxic Emissions

Exposure to hazardous chemicals is a concern for workers in fossil fuel facilities. Robust safety protocols, including the use of personal protective equipment (PPE), adequate ventilation, and regular monitoring of air quality, help minimize the risk of chemical exposure. Additionally, facilities implement measures to control and reduce toxic emissions to the environment.

9.5.3.3. Process Safety Management

Fossil fuel facilities implement Process Safety Management (PSM) systems to identify, assess, and mitigate potential risks associated with the manufacturing processes. This involves comprehensive risk assessments, safety audits, and the development of protocols to handle abnormal operations and emergencies.

9.5.3.4. Pressure Vessel and Equipment Safety

Pressure vessels, pipelines, and other critical equipment are integral components of fossil fuel facilities. Regular inspections, maintenance, and adherence to strict engineering standards are essential to ensure the integrity and safety of these components.

9.5.3.5. Occupational Health and Safety

Worker safety is paramount in fossil fuel facilities. Training programs, safety drills, and the use of advanced automation and remote monitoring technologies contribute to minimizing direct exposure to hazardous environments and reducing the risk of accidents.

9.5.3.6. Environmental Impact and Spill Response

Fossil fuel facilities are under scrutiny for their environmental impact, including the potential for oil spills or chemical releases. Comprehensive spill response plans, containment systems, and monitoring initiatives are in place to minimize environmental damage and facilitate swift cleanup in the event of a spill.

9.5.3.7. Security Measures

Given the critical nature of fossil fuel facilities and their importance to the energy infrastructure, security measures are implemented to protect against potential threats, including terrorism or sabotage. These measures may include surveillance systems, access controls, and emergency response protocols.

9.5.3.8. Community Outreach and Emergency Planning

Fossil fuel facilities engage with local communities to share information about safety measures, emergency response plans, and potential risks. Collaborative efforts between facility operators, emergency responders, and community stakeholders contribute to effective emergency preparedness and response.

9.5.4. Future Trends and Innovations

The future of renewable energy is poised for transformative advancements driven by technological innovations, policy developments, and evolving market dynamics. As the world increasingly shifts towards sustainable energy solutions, several key trends and innovations are set to redefine the landscape of renewable energy.

1. **Advanced Energy Storage:** One of the most significant trends is the evolution of energy storage technologies. While lithium-ion batteries remain dominant, innovations such as solid-state batteries, flow batteries, and metal-air batteries are emerging, offering higher energy densities, improved safety, and longer life spans. These advancements will enhance the ability to store large quantities of renewable energy and provide reliable power during periods of low generation, thus addressing the intermittency of sources like wind and solar.
2. **Green Hydrogen:** Green hydrogen, produced through the electrolysis of water using renewable energy, is gaining momentum as a versatile and sustainable energy carrier. It offers a promising solution for sectors difficult to decarbonize with electricity alone, such as heavy industry, transportation, and long-term energy storage. As electrolysis technology becomes more efficient and cost-effective, green hydrogen is expected to play a crucial role in achieving net-zero emissions.
3. **Smart Grids and Digitalization:** The integration of smart grid technologies and digital tools is transforming how energy is managed and distributed. Smart grids use advanced sensors, real-time data analytics, and automation to enhance grid reliability, optimize energy flows, and integrate diverse renewable energy sources. Digital platforms and artificial intelligence (AI) are also being employed to predict energy demand, manage distributed energy resources, and improve overall grid efficiency.
4. **Floating Solar and Offshore Wind:** Expanding beyond traditional land-based installations, floating solar panels and offshore wind farms represent significant growth areas. Floating solar systems can be deployed on bodies of water where land is scarce or expensive, while offshore wind farms take advantage of stronger and more consistent winds found at sea. Both technologies are expected to scale rapidly and contribute substantially to global renewable energy capacity.

5. **Energy Efficiency and Demand Response:** Innovations in energy efficiency, such as advanced building materials, smart appliances, and high-performance HVAC systems, are reducing overall energy consumption. Demand response programs, which incentivize consumers to adjust their energy usage based on grid conditions, are also becoming more sophisticated, helping to balance supply and demand and integrate higher shares of renewables and integrate higher shares of renewables.
6. **Circular Economy and Recycling:** The renewable energy sector is increasingly focusing on the circular economy, emphasizing the recycling and repurposing of materials used in renewable technologies. Innovations in the recycling of solar panels and wind turbine blades are improving sustainability and reducing the environmental impact of renewable energy systems.

In conclusion, the future of renewable energy is marked by rapid technological advancements and innovative solutions that enhance efficiency, reliability, and scalability. These trends will play a critical role in accelerating the transition to a low-carbon energy system and achieving global sustainability goals.

9.5.5. Next-Generation Renewable Energy Technologies

The energy landscape is undergoing a transformation led by next-generation renewable energy technologies, which offer improved sustainability, scalability, and efficiency. Emerging trends in renewable energy present both opportunities and challenges for innovation, placing the renewable energy sector at a pivotal and exciting crossroads. In the face of global challenges like climate change and depleting natural resources, this industry comes out as a ray of hope, presenting workable solutions along with a wealth of new technological developments and expanding investment opportunities.

9.5.5.1. Sustainable Energy Technological Innovations

1. **Green Hydrogen's Emergence:** Green hydrogen—which is generated from renewable resources—is anticipated to play a bigger role in the energy economy by 2024. We expect significant developments in electrolyzer technology, which will lower costs and increase efficiency in the production of green hydrogen. These advancements are critical because they will increase the viability of green hydrogen in a number of areas, such as transportation and industrial processes. This development underscores the potential

of green hydrogen in the larger shift to sustainable energy systems and represents a significant step toward diversifying applications of renewable energy.

2. **Solar and Wind Energy Advancements:** Solar and wind energy are ready for revolutionary improvements. The tip of the iceberg includes massive offshore wind turbines with gigawatt capacities and bifurcated solar panels that receive sunlight from both sides. These developments will make it possible to harvest renewable resources more effectively, increasing their viability and ability to compete with more conventional energy sources like wind and solar power. One noteworthy trend is the increase in solar PV capacity, which is expected to account for a large amount of the growth in renewable energy sources and surpass one terawatt of global solar power generation. Furthermore, it is anticipated that the percentage of offshore wind energy will rise as more nations, such as the US, Canada, China, India, and the UK, develop their offshore capacities.
3. **Breakthroughs in Energy Storage:** Developments in energy storage technologies will have a significant impact on the expansion of renewable energy in 2024. Innovative storage options like solid-state batteries, which have longer lifespans and higher energy densities than conventional lithium-ion batteries, are probably going to become more prevalent. It's also anticipated that advancements in gravity-based storage technologies will pick up steam. The long-term, sustainable storage options offered by these systems are crucial for controlling the intermittent nature of renewable energy sources. These developments are essential to guaranteeing a steady and dependable supply of renewable energy.

9.5.5.2. Sustainable Energy Policy Dynamics

1. **Enhanced Government Support:** It is anticipated that governments will be essential in hastening the switch to renewable energy. Subsidies, tax breaks, and regulatory frameworks that support renewable energy projects may grow in number by 2024. These regulations will be crucial for promoting the use of renewable energy sources and drawing in capital. China is still at the forefront of renewable energy, hoping to surpass its goal of obtaining 33% of its electricity from renewable sources by 2025 thanks to notable advancements in wind power.
2. **Grid Integration Initiatives:** One of the main areas of focus in 2024 will be the integration of renewable energy into the current power

grids. The focus will be on modernizing the grid's infrastructure to efficiently handle the variability of renewable energy sources. This will involve putting cutting-edge technologies like distributed energy systems and smart grids into practice, which can improve the distribution and dependability of energy. These programs are necessary to guarantee a reliable and effective energy system that can manage a higher proportion of renewable energy, allowing for a more seamless transition to a more sustainable energy future.

9.5.5.3. Renewable Energy Investment Trends

1. **Corporate Investment Surge:** In 2024, it is anticipated that corporate investment in renewable energy will increase dramatically. Leading businesses in important industries like food and beverage, textile, pharmaceutical, automotive, and logistics are investing in green energy projects and setting goals for renewable energy as part of their sustainability strategies. The growing understanding of the long-term financial returns and corporate responsibility benefits of renewable energy investments is what is driving this trend. New renewable projects may face difficulties due to the rising costs of materials, which are influenced by the dynamics of the world market. However, the market has maintained the competitiveness of renewable energy due to the rising cost of fossil fuels and the requirement for energy security.
2. **Expansion of Green Bonds and ESG Investments:** It is projected that in 2024, the Environmental, Social, and Governance (ESG) and green bond markets will grow significantly. This pattern points to a larger movement in the investing community in the direction of social responsibility and sustainability. A growing number of investors are looking for opportunities that not only yield financial gains but also make a positive impact on societal and environmental objectives. More cutting-edge financial products in the green bond and ESG sectors are probably going to result from this growing interest, giving investors a variety of ways to interact with and support sustainable projects around the globe.

9.5.5.4. Emerging Energy Market Dynamics

1. **Developing Countries Leaping Forward:** It is anticipated that developing nations will quickly switch to renewable energy, eschewing the infrastructures based on conventional fossil fuels. Growing international support for sustainable development

initiatives, coupled with the decreasing costs and increasing accessibility of renewable technologies, are the driving forces behind this trend. The use of renewable energy in these areas has the potential to drastically change the global energy scene and create new chances for development and innovation. To combat climate change, nations throughout the world are looking to invest in affordable renewable energy, with an emphasis on lowering dependency on imported energy, particularly in the U.S. and Europe.

2. **Decentralized Renewable Energy Systems:** Decentralized renewable energy systems are expected to proliferate in 2024, particularly in isolated and rural areas. These systems improve local resilience and lessen dependency on major infrastructure by providing a dependable and sustainable substitute for conventional grid-based energy. Furthermore, it is anticipated that the transportation industry will continue to be a major driver of the growing demand for biofuels. Supportive government policies in major international economies such as the US, Brazil, Europe, India, and Indonesia have contributed to this growth, indicating a wider commitment to diversifying energy sources and promoting sustainable fuel alternatives.

9.5.5.5. Consumer Trends and Awareness

1. **Growing Consumer Demand:** The need for renewable energy will probably increase and consumer awareness will rise more sharply in 2024. propelled by an increase in sustainable living methods and a greater understanding of the effects of climate change. This change is resulting in better informed consumer choices, with the preference for renewable and environmentally friendly energy sources. This trend is being further amplified by social media and influencer campaigns, which raise awareness and promote sustainable lifestyle choices. As a result, consumer preferences will have a greater impact on market dynamics, driving businesses to use greener technologies and processes and pushing governments to enact more laws that will support the use of renewable energy sources.

2. **Role of Digital Platforms:** Customers will have more control over how much energy they use with the introduction of smart home technologies and Internet of Things devices. It is projected that more people will use apps and platforms that give real-time data on energy usage, renewable energy alternatives, and even gamification features to promote energy-saving behaviors. These digital tools

help utility companies engage with customers directly and build a more engaged and energy-conscious community by empowering consumers to make better-informed decisions. AI-driven suggestions for maximizing energy use will probably be added to these platforms in the future, further encouraging sustainability and efficiency.

9.5.6. Vision for the Future: Long-Term Strategies for Achieving Significant Carbon Footprint Reduction through Renewable Energy

Achieving a significant reduction in carbon footprint through renewable energy requires a multifaceted, long-term strategy that integrates technological advancements, policy frameworks, and systemic changes.

Here's a vision for the future that outlines key strategies for realizing this goal:

1. Accelerating Technological Innovation
    - **Investment in R&D:** Continuous research and development in renewable energy technologies are crucial. This includes advancing solar photovoltaics, wind turbines, energy storage systems, and emerging technologies like green hydrogen and advanced geothermal systems. Increased funding and collaboration between public and private sectors can drive breakthroughs that enhance efficiency, lower costs, and expand application areas.
    - **Deployment of Next-Generation Technologies:** Rapidly scaling up next-generation technologies such as solid-state batteries, floating solar panels, and floating wind turbines can significantly boost renewable energy capacity. Supporting pilot projects and early adoption can help demonstrate their viability and attract further investment.
2. Expanding Renewable Energy Infrastructure
    - **Grid Modernization:** Upgrading the electricity grid to a smart grid that can handle decentralized and variable renewable energy sources is essential. This involves enhancing transmission lines, developing energy storage solutions, and implementing advanced grid management technologies to improve reliability and efficiency.
    - **Infrastructure Development:** Expanding and diversifying renewable energy infrastructure, including large-scale solar

farms, offshore wind farms, and community-based projects, can increase renewable energy generation and reduce reliance on fossil fuels. Strategic site selection and integration of diverse renewable sources can optimize energy production.

3. Supporting Policy and Regulatory Frameworks
   - **Comprehensive Climate Policies:** Governments should implement robust climate policies and regulations that support renewable energy adoption. This includes setting ambitious renewable energy targets, providing incentives for clean energy investments, and establishing carbon pricing mechanisms to internalize the environmental costs of fossil fuels.
   - **Long-Term Energy Planning:** Creating and adhering to long-term energy plans that prioritize renewable energy integration and carbon reduction goals is crucial. Policies should encourage innovation, support renewable energy research, and facilitate the transition from fossil fuels to cleaner energy sources.
4. Enhancing Energy Efficiency
   - **Efficiency Standards:** Implementing and enforcing energy efficiency standards for buildings, appliances, and industrial processes can reduce overall energy consumption and complement renewable energy efforts. Efficient energy use reduces demand and maximizes the impact of renewable energy generation.
   - **Demand Response Programs:** Developing and expanding demand response programs can help balance energy supply and demand, integrating higher shares of renewable energy. By incentivizing shifts in energy use patterns, these programs can improve grid stability and reduce the need for backup fossil fuel power.
5. Promoting Decarbonization Across Sectors
   - **Electrification of Transportation:** Transitioning to electric vehicles (EVs) and expanding EV infrastructure can reduce emissions from the transportation sector. Integrating EVs with renewable energy sources and smart grids can further enhance their environmental benefits.
   - **Industrial Decarbonization:** Supporting the adoption of renewable energy and clean technologies in industrial

processes can reduce emissions from manufacturing and heavy industries. This includes promoting energy efficiency, green hydrogen use, and carbon capture and storage (CCS) technologies.

6. Fostering Global Collaboration
   - **International Agreements:** Engaging in global climate agreements and initiatives can drive collective action towards renewable energy adoption and carbon reduction. International cooperation on technology sharing, funding, and best practices can accelerate progress and ensure a unified approach to climate challenges.
   - **Knowledge Exchange:** Promoting knowledge exchange and capacity building among countries can help disseminate successful renewable energy practices and technologies, fostering a global transition to sustainable energy systems.
7. Engaging Communities and Stakeholders
   - **Public Awareness:** Raising awareness about the benefits of renewable energy and the importance of reducing carbon footprints can foster public support and drive behavioral change. Education campaigns and community engagement are essential for building a culture of sustainability.
   - **Stakeholder Collaboration:** Engaging stakeholders, including businesses, local governments, and non-governmental organizations, in the planning and implementation of renewable energy projects ensures diverse perspectives and effective solutions. Collaborative efforts can enhance project success and support broader climate goals.

Achieving significant carbon footprint reduction through renewable energy requires a comprehensive and coordinated approach. By advancing technology, expanding infrastructure, supporting effective policies, enhancing energy efficiency, and fostering global and local collaboration, we can create a sustainable energy future that addresses climate change and promotes environmental stewardship (Buonocore et al., 2015; Cacciapaglia et al., 2018).

REFERENCES

1. Abbasi, T., & Abbasi, S. A. (2011). 'Renewable' hydrogen: Prospects and challenges. *Renewable and Sustainable Energy Reviews, 15*, 3034–3040.
2. Abdalla, M. A., Hossain, S., Nisfindy, O. B., Azad, A. T., Dawood, M., & Azad, K. A. (2018). Hydrogen production, storage, transportation and key challenges with applications: A review. *Energy Conversion and Management, 165*, 602–627.
3. Akal, D., Öztuna, S., & Büyükakin, M. K. (2020). A review of hydrogen usage in internal combustion engines (gasoline-LPG-diesel) from combustion performance aspect. *International Journal of Hydrogen Energy, 45*, 35257–35268.
4. Alonso, E., & Field, F. R. (2018). Environmental, health, and safety issues in petroleum refineries. In *Sustainable Oil Industry* (pp. 75–100). Springer.
5. Baird, A. R., Archibald, E. J., Marr, K. C., & Ezekoye, O. A. (2020). Explosion hazards from lithium-ion battery vent gas. *Journal of Power Sources, 446*, 227257.
6. Boyle, A. D., Leggat, G., Morikawa, L., Pappas, Y., & Stephens, J. C. (2021). Green New Deal proposals: Comparing emerging transformational climate policies at multiple scales. *Energy Research & Social Science, 81*, 102259.
7. Buonocore, J. J., et al. (2015). Health and climate benefits of different energy-efficiency and renewable energy choices. *Nature Climate Change*.
8. Byrd, J. M., & Wachsman, E. D. (2015). The role of solid oxide fuel cells in the safety and reliability of hydrogen infrastructure. *Energy & Environmental Science, 8*(10), 2866–2881.
9. Cacciapaglia, A., Saldaña, D. A., & Wei, S. (2018). Cybersecurity risks in the oil and gas sector. *Energy Reports, 4*, 155–161.
10. Chandler, H. D. (2018). *Guidelines for the Safe Use of Hydrogen in the Oil and Gas Industry*. Wiley.
11. Christensen, P. A., Anderson, P. A., Harper, G. D. J., Lambert, S. M., Mrozik, W., Rajaeifar, M. A., Wise, M. S., & Heindrich, O. (2021). Risk management over the life cycle of lithium-ion batteries in electric vehicles. *Renewable and Sustainable Energy Reviews, 148*, 111240.
12. Hester, R. E., & Harrison, R. M. (2017). *Risk Analysis and Control for Industrial Processes—Gas, Oil and Chemicals: A System Perspective for Assessing and Avoiding Low-Probability, High-Consequence Events*. Royal Society of Chemistry.
13. Huang, B., & Liu, Z. (2017). Safety in nuclear power plants. In *Handbook of Nuclear Engineering* (pp. 1–33). Springer.

14. Kuznetsov, V. V., & Makarov, A. A. (2016). Safety issues of hydrogen energy use. *International Journal of Hydrogen Energy, 41*(33), 14692–14702.

15. Lees, F. P. (2013). *Lees' Loss Prevention in the Process Industries: Hazard Identification, Assessment and Control*. Elsevier.

16. Mohammadi, M., & Ahmadi, A. (2015). *Natural Gas Processing from Midstream to Downstream: Vision, Strategy, and Execution*. Gulf Professional Publishing.

17. Organization of the Petroleum Exporting Countries (OPEC). (2019). *Oil and Gas Safety Performance: A Statistical Review*.

18. Ortega, E., & Zhang, W. (2017). Critical review on the safety issues of hydrofluoric acid and its derivatives in the petroleum industry. *Journal of Loss Prevention in the Process Industries, 49*, 32–43.

19. Rausand, M., & Høyland, A. (2014). *System Reliability Theory: Models, Statistical Methods, and Applications*. Wiley.

20. Renewable Energy Policy Network for the 21st Century (REN21). (2020). *Renewables 2020 Global Status Report*.

21. Seffen, K. A. (2017). Safety of nuclear power plants. In *Advances in Human Factors and Ergonomics* (pp. 439–448). CRC Press.

22. United Nations Industrial Development Organization (UNIDO). (2017). *Biomass Energy for Industry in Developing Countries: A Review of The Status and Prospects for The Adoption of Renewable Energy in The Industrial Sector*.

INDEX

D

E

T

U

V

W

Printed in the United States
by Baker & Taylor Publisher Services